THE SHORTER
SCIENCE AND CIVILISATION IN CHINA

This is an authorized Taiwan Edition
published under special agreement with
the Proprietor for sale in Taiwan only.

First Printing.....July 1986

Published by

CAVES BOOKS, LTD.

103, Chungshan N. Rd., Sec. 2, Taipei
Tel: 5414754, 5371666

台灣版

有著作權◇不准翻印

台內著字第　　　　　號

發行人：周　　　　政
發行所：敦煌書局股份有限公司
地　址：台北市中山北路2段103號
電　話：(02) 537·1666（總機）
郵　撥：００１４１０３－１
印刷所：聯和印製廠有限公司
新聞局登記證局版台業字0269號
中華民國75年 7 月　　日第 版

定價　NT$

COLIN A. RONAN
Project Co-ordinator, East Asian History of Science Trust

The Shorter Science and Civilisation in China

AN ABRIDGEMENT OF
JOSEPH NEEDHAM'S ORIGINAL TEXT

Volume 3

A SECTION OF VOLUME IV, PART 1
AND A SECTION OF VOLUME IV, PART 3
OF THE MAJOR SERIES

Cambridge University Press

CAMBRIDGE

LONDON NEW YORK NEW ROCHELLE

MELBOURNE SYDNEY

Published by the Press Syndicate of the University of Cambridge
The Pitt Building, Trumpington Street, Cambridge CB2 1RP
32 East 57th Street, New York, NY 10022, USA
10 Stamford Road, Oakleigh, Melbourne 3166, Australia

First published 1986.

Library of Congress catalogue card number: 77-82513

British Library Cataloguing in Publication Data

Needham, Joseph
The Shorter Science and Civilisation in China:
an abridgement of Joseph Needham's original text.
Vol. 3

1. Science – China – History
2. Technology – China – History
I. Title   II. Ronan, Colin A.
509' 51   Q127.C5

ISBN 0 521 25272 5 hard covers
ISBN 0 521 31560 3 paperback

CONTENTS

ILLUSTRATIONS

TABLES

PREFACE

In this, the third volume of the abridgement of Dr Joseph Needham's *Science and Civilisation in China* we look in some detail at the greatest contribution the Chinese made to physics – the discovery of the magnetic compass. We delve into its origins in divination and study its links with a form of proto-chess, then pass on to its application in navigation. This brings us to the fascinating subject of Chinese ships and shipping, and the voyages made to the West and even further afield at times that predate equivalent voyages from the Middle East or by mariners from the West. This volume therefore covers what is the last section of volume IV, part 1 of *Science and Civilisation in China* and the second half of volume IV, part 3. In doing so the abridgement does not therefore follow precisely the order of subjects in Dr Needham's original text, but it was felt that to depart from that arrangement would be more satisfactory from the point of view of the reader of this shortened version, who may wish to have certain related subjects grouped together in a single volume.

Yet again I have received much encouragement and help from Joseph Needham, whose advice has been invaluable and who has once more generously given precious time to preparing the bibliography. As in the previous two volumes, this is no new edition. However, it has seemed desirable in view of the increasingly widespread use of the Pinyin romanisation of Chinese to insert that system in square brackets after the modified Wade–Giles system used throughout the main work and in the previous two volumes of this abridgement. For reasons which will become obvious, Pinyin is not given on every occasion, though it appears usually once in every paragraph in which a romanised Chinese word appears in the modified Wade–Giles form. Only at those rare times when the romanisation is the same in both systems is no Pinyin equivalent given at all.

My warmest thanks are due to Commander Henry Hatfield, RN (ret'd) for reading through the nautical sections, to Professor David Davies and Susannah Perry for help with the Index, and to Dr Simon Mitton of the Press for his patience and my copy editor for his care.

Bar Hill, Cambridge   Colin A. Ronan
August 1984

Magnetism and electricity

With this volume we arrive at what was the greatest Chinese contribution to physics, the discovery of magnetism. It is a subject that has given rise to a vast literature because the attractive power of the lodestone was known, both in China and the West, from at least the middle of the first millennium BC. There is no doubt about this, but there is doubt about the discovery of the lodestone itself and of the fact that pieces of iron could become magnetised by contact with it. Knowledge of this appeared rather suddenly in Europe at the very end of the twelfth century AD, and search for immediate antecedents in Arabia and India has not proved successful.

That the Chinese were the first to understand and utilise the directive property of the lodestone has traditionally been admitted, but strangely enough for entirely the wrong reasons. From Han times (late third century BC) onwards Chinese texts speak of the 'south-pointing carriage' (*ting nan chhê* [*ding nan che*] or *chih nan chhê* [*zhi nan che*]), the art of making which was continually being lost and revived. From the time of the Jesuit missionaries in the early seventeenth century onwards, it was assumed that these were references to some form of magnetic compass, but it may now be considered solidly established that the south-pointing carriage had nothing whatever to do with magnetic directions. It was essentially a self-regulating mechanical device with a system of gear-wheels to ensure that a pointer always pointed to the south, irrespective of the motion of the carriage. Some contemporary historians of science, unaware of this, have criticised Chinese literature for what they have taken to be claims for discovery, whereas what they have come across are legends concerning the south-pointing carriage. Moreover, as we have had occasion to remark before in other volumes of this abridgement, some historians of science have even gone so far as to doubt whether anything having a bearing on Western science could have come from China. Yet none have been able to discover any precursors of European knowledge of the lodestone's direction-seeking power before the turning point of AD 1190.

The story of the magnetic compass in China was revolutionised in the late 1940s and early 1950s by Wang Chen-To [Wang Chen-Duo] who managed to explain a fundamental text in the *Lun Hêng* [*Lun Heng*] (Discourses Weighed in the Balance) of AD 83. This revealed a probable connection between the magnetic compass and the diviner's board of the Han people. In what follows we shall seek to show (*a*) that the first text clearly describing the magnetic needle compass is undeniably of about AD 1080, i.e. a century earlier than the first European mention of this instrument; (*b*) that magnetic declination (i.e. the failure of the magnetic needle to point to true or geographical north), as well as the magnet's directivity, is mentioned there; and (*c*) that the declination was discovered in China some time between the seventh and tenth centuries AD. It will also become evident, (*d*) that the use of the needle, which alone permitted the construction of an accurate pointer-reading instrument, was the limiting factor in discovering declination, and belongs to the beginning of this period; and (*e*) that the original Chinese compass was probably a kind of spoon carefully carved from lodestone and revolving on the smooth surface of a diviner's board. Lastly we shall arrive at Dr Needham's suggestion that there is a detectable connection not only with divination practices but also with games such as chess. The original form, the lodestone spoon, was certainly known and used in the first century AD, and may go back, as a secret of court magicians, two centuries earlier. The failure to elucidate this before has been partly due to scholars searching for traces of a south-pointing vehicle when they ought to have been looking for a 'south-controlling spoon' (*ssu nan shao*) [*si nan shao*].

The case of the earliest magnetic compass in navigation is somewhat similar. It is certain that by AD 1190 it was in use in the Mediterranean, but its use is also spoken of in a Chinese text just under a century previously. A mistranslation of this text by earlier scholars led to the persistent statement that it was then found only on foreign (Arab) ships trading to Canton [Guangzhou], but this idea, as we shall see, has no basis. Something will be said later as to possible means of transmission; the famous Elizabethan investigator of terrestrial magnetism, Wiliam Gilbert, thought that Marco Polo or someone else of his time brought it, but this would have been a century too late. 'If the Chinese', wrote the historian Edward Gibbon, 'with the knowledge of the compass, had possessed the genius of the Greeks and Phoenicians, they might have spread their discoveries over the southern hemisphere'. In fact, this is just what they did.

MAGNETIC ATTRACTION

What was known of magnetic attraction in the medieval West? By the beginning of the Middle Ages it had been established (*a*) that the lodestone attracts pieces of iron; (*b*) that it does so across a distance; (*c*) the attracted iron adheres to the magnet; (*d*) the magnet induces a power of attraction in the attracted iron, which (*e*) it retains for some time. It had also been observed that (*f*) the magnetic influence would act through substances other than iron, and (*g*) some magnets would repel pieces of iron as well as attract them. Tradition by way of Aristotle tells us that Thales (sixth century BC) studied the magnet using animistic explanations based on animal or human attraction, while another tradition refers to later philosophers taking an interest in the subject, and certainly Democritus (fifth century BC) discussed it. Lastly, all the fundamental properties mentioned above were described in the first century BC by Lucretius, who suggested that attraction was due to a vacuum being established between the iron and the magnet.

In China, as in Europe, the magnet went by many names. The most usual was *tzhu shih* [*ci shi*], the 'loving stone', generally combined in the ideograph *tzhu* [*ci*] (磁). Another derivative *tzhu* [*ci*] (孳) means copulation or breeding, and those who first observed magnetism in China seem, like Thales, to have thought of attraction animistically. Again, *hsûan shih* [*xuan shi*], the 'mysterious stone', though it later signified non-magnetic iron ore, seems likely, in view of other phrases involving *shih* [*shi*], to have originally referred to the lodestone. Most of these and other names and phrases go back to the Chin [Jin] (second to fifth centuries AD) or at least to the Thang [Tang] (seventh to tenth centuries).

Chinese literature between the third century BC and the sixth century AD is as full of references to the attractive power of the magnet as the European. Though there is nothing as early as Thales, nor contemporary with Aristotle, the *Lü Shih Chhun Chhiu* [*Lü Shi Chun Qiu*] (Master Lü's Spring and Autumn Annals) – a compendium of science which mentions attraction – would be late third century BC, about contemporary with Archimedes. And shortly before Lucretius was writing, the *Huai Nan Tzu* [*Huai Nan Zi*] (The Book of (the Prince of) Huai-Nan) – another compendium of science – said:

> If you think that because the lodestone can attract iron you can also make it attract pieces of pottery, you will find yourself mistaken....
> The lodestone can attract iron but has no effect on copper. Such is the motion of the Tao [Dao].

Elsewhere the writer also says that 'the lodestone flies upwards', meaning that a small piece of magnetite could be attracted by iron held above it. By AD 83 the *Discourses Weighed in the Balance* was commenting that both magnetic attraction and the attraction exhibited by amber were examples of 'sympathetically attracting things'. Wang Chhung [Wang Chong] who wrote the book, also refers to the 'mutual influence' displayed by amber and the lodestone on various bodies, remarks which echo the general Chinese concept of 'resonance' and action at a distance.

In the fifth century AD the Chinese had already begun to undertake measurements of magnetic force. The medicinal properties of lodestone were considered to be different from those of non-magnetic iron ore, and it was necessary to distinguish between the two, particularly as the latter was sometimes rather toxic. So in the *Lei Kung Phao Chih* [*Lei Gong Pao Zhi*] ((Handbook based on the) Venerable Master Lei's (Treatise on) the Preparation (of Drugs)) we read:

> If you want to make a test take one catty (about 600 gm) of the stone and see whether using all four sides it can attract an equal weight of pieces of iron – if so this is the best, and may be called *yen nien sha* [*yan nian sha*]. The sort which (in the same conditions) on all four faces attracts 200 gm is called *hsü tshai shih* [*xu cai shi*]. Again that which will only attract about 100 or 140 gm is termed (ordinary) lodestone (*tzhu shih* [ci shi]).

Pesumably stones of less power were graded into the category of non-magnetic ore. The method of estimation, involving as it did the use of the balance cannot be later than the twelfth century AD, texts of which often quote it, and may well be more than five hundred years earlier.

In both East and West numerous legends grew up about the lodestone. They took various forms: there were certain islands which ships could not pass if they were constructed with iron nails, or gates which men could not get through if armed with iron weapons; alternatively it was thought that somewhere or other statues of iron floated in mid-air, suspended by magnetic attraction. In the second century AD Ptolemy, the Greek astronomer and geographer, placed such magnetic islands between Ceylon and Malaya, while we find the same story in the *Nan Chou I Wu Chih* [*Nan Zhou Yi Wu Zhi*] (Strange things of the South) two centuries later. But probably the idea had its own forms of purely Chinese origin, for a palace at Chhang-an (Sian) was long supposed to have magnetic gates to deter invaders, while such gates were also connected with myths about ordeals and escape from the mundane world. Similar stories occur in Arabic texts.

It was natural that the lodestone should find application both in alchemy and medicine. Sung [Song] medical books (fifth century AD) often speak of opening blocked passages or extracting foreign bodies such as needles or arrow fragments by the use of lodestones – which suggests that even if the processes were more imaginative than successful, it was clearly realised that magnetic attraction acted through intervening substances other than iron. Magnets only became much used in European medicine in the seventeenth and eighteenth centuries AD.

On the whole it may be said that between ancient and medieval knowledge of attraction in Europe and China there was nothing to choose. One finds less theory about it in China, perhaps because action at a distance fitted in better with the Chinese world-view than with the Greek. Indeed, in Greece things were thought to have their 'natural places' in the cosmos and magnetic behaviour was difficult to reconcile with this concept, as also with Aristotle's doctrines of the 'natural' and violent' motion of bodies.

Nevertheless a particulary Chinese conception was due to the Greek heretic Hermogenes (second century AD), who believed that God had created the world from nothing, but had organised all matter by acting upon it like a magnet. If this patterning principle had been brought down to earth as a physical reality, instead of being transcendent, there would have been no difference between it and the Chinese doctrine of the Tao [Dao].

Lastly we may note that while Pliny (first century AD) and later writers in the West referred to the fact that sometimes the lodestone would repel iron, the Chinese were also aware of this. There is a curious story of magnetised chess-men (p. 47) which has come down to us in several versions, while words meaning 'push away' and 'repel' were also used and, in the first century BC, the *Shih Chi* [*Shi Ji*] (Memoirs of the Historiographer(-Royal) down to 99 BC) says the pieces 'mutually hit' each other. We have then a hint that in China repulsion was also observed.

ELECTROSTATICS

Just as ancient and medieval Chinese knowledge of the lodestone paralleled that of the West, so also with the fact that certain substances like amber will, when rubbed, acquire the ability to attract small objects such as dried plant fragments. It is again Thales who is credited with the first of such observations, and whereas the electrum of Homer (ninth or eighth century BC) was an alloy of gold and silver, certainly after the time of Herodotus (fifth century BC) the word generally refers to amber, and our 'electricity' derives from it. Plato mentioned it, but only Plutarch and Pliny (first century AD) state that it must be rubbed beforehand.

Greek amber was probably of Baltic origin, but most Chinese amber came from deposits of burmite in Upper Burma. This has slightly different chemical and physical properties from succinite, the common amber, though both are the fossil gum from conifers. Wang Chhung [Wang Chong], a contemporary of Plutarch and Pliny, is one of the first to mention it; he used the term *tun mou* [*dun mou*] which is likely to have been formed from some Shan or Thai language. Such is also supposed to be the derivation of the more usual term *hu pho* [*hu po*]. About AD 500, the physician and alchemist Thao Hung-Ching [Tao Hong-Jing] wrote that amber was fir-tree resin which had been buried for a thousand years, and in which entrapped insects might be seen, adding that there was a method of imitating it by heating hen's eggs with dark fish roe. But, he points out,

> ... only that kind which, when rubbed with the palm of the hand, and thus made warm, attracts mustard-seeds, is genuine.

The electrostatic test for genuine amber is still used. And though Wang Chhung had not mentioned the need for rubbing, after the Han practically all pharmacopoeias mention amber and its properties. After this there was no real advance in China, any more than in Europe, until the study of electricity really began in the eighteenth century.

MAGNETIC DIRECTIVITY AND POLARITY

The magnetic compass is the oldest representative of all those dials and pointer-readings which play so great a part in modern science. The sun-dial was of course far older, but there only a shadow moved, not part of the instrument itself. The wind-vane was older, but there the possibility of precise readings on a circular graduated scale was absent in all its ancient forms. The magnetic compass was also self-registering. Furthermore, its development, in which a needle allowing accurate readings replaced a piece of lodestone, and the later use of pointers for precise indications on a scale, displays the process of induction – reasoning from the particular to the general. No apology is needed, therefore, to consider in some detail what was the oldest form of compass developed by the Chinese, and when it was that successive developments were introduced. That so fundamental an instrument spread so slowly is not difficult to understand once we realise that its original discovery was connected with a divination process by imperial magicians, and that since it then developed in an agrarian rather than a primarily maritime civilisation, its use was for centuries limited to a specifically Chinese pseudo-science, Taoist [Daoist] geomancy, details of which reached a high level of refinement. The adoption of the compass by Chinese sailors was probably long retarded also because all through the Middle Ages, river and canal traffic predominated over ocean voyages.

Geomancy was mentioned in volume I of this abridgement, but a brief reference to its main points may be desirable all the same. For the Chinese the term meant 'the art of adapting residences of the living and tombs for the dead so as to co-operate and harmonise with the local currents of the cosmic breath'. Known as the science of 'winds and waters' (*fêng shui* [*feng shui*]), it referred not merely to physical winds but rather the *chhi* [*qi*] or *pneuma* (spirit) of the earth circulating through the veins and vessels of the earthly macrocosm. The waters too were not only the visible streams and rivers but also those passing to and fro out of sight, removing impurities, depositing minerals, and like the *chhi*, affecting for good or evil the houses and families of the living, as also the descendants of those who lay in the tombs. The history of the magnetic compass is only understandable in the context of this system of ideas, for this is the matrix in which it was generated.

Of all forms of divination, geomancy was perhaps that which became most deeply rooted in Chinese culture throughout the traditional period. It led to a minute appreciation of topographical features, because the protection of a site from harmful influences was always a matter of great importance. Purely superstitious in many respects though the subject sometimes became, the system as a whole undoubtedly contributed to the exceptional beauty of positioning of buildings, villages and cities throughout the realm of Chinese civilisation (Fig. 166).

The system seems certainly to have developed during the fourth century BC in the Warring States period when the natural philosopher Tsou Yen [Zou Yan] and schools of philosophical magic were flourishing. Thus the *Kuan Tzu* [*Guan Zi*] (The Book of Master Kuan) speaks of water as the blood and breath of the earth, 'flowing and communicating within its body as if in sinews and veins'. And a century later, just before his death in 210 BC, Mêng Thien [Meng Tian], the builder of the Great Wall, claimed that he could not have built it 'without cutting through the veins of the earth'. In the Han then, the system was well under way, and was consolidated during the San Kuo [San Guo] period (AD 221–65).

From the Thang [Tang] (seventh to ninth centuries) onwards, the rise of the compass seems to have led to a division of the geomancers into two schools. Those who came from Chiangsi [Jiangxi] province held fast to many of the older principles, reasoning their way on the shapes of mountains and the courses of rivers, no doubt much as their Han predecessors had done. On the other hand, the men of the maritime region of Fukien [Fujian] regarded the compass as all important for determining changes in topography, though besides this they also made much use of the symbols in the *Book of Changes* (see volume 1 of this abridgement) and paid more attention to astrology. The marks of this division are still evident in Ming and Chhing [Qing] works.

The surviving literature on geomancy and the compass is quite large, but even so there are still gaps and obscurities. Indeed, it is to be feared that some of the most interesting facts about the development of one of the most important of all scientific instruments have perished for ever. Some items, it seems, were wilfully destroyed, for while there may be a much doubt about the traditional burning of books attributed to the first emperor Chhin Shih Huang Ti [Qin Shi Huang Di], there is clear evidence for destruction among Jesuit converts in the seventeenth century. This was tragic and paradoxical in view of the learned nature of the Jesuit mission.

唐朝敕賜狀元鄒應龍祠形與地形之圖

Fig. 166. Geomancy as the background to the science of magnetism: a selected layout of buildings and environment in the early thirteenth century AD. The caption at the top says: 'Map of the Grounds and Family Temple donated by Imperial Rescript to Tsou Ying-Lung [Zou Ying-Long], Optimus Graduate in the Palace Examinations'. In a region of hills three streams join together at a point left of centre to form a small river, which flows away at the left bottom corner. The temple, backed by two small lakes, faces the upper part of the plan, with a view of the auspicious point (marked by a pavilion) at the tip of the hilly ridge separating the two upper valleys, each of which contains numerous rice-fields. Two bridges and a gateway in the hills are also marked. From *Ti Li Cho Yü Fu* [*Di Li Zhou Yu Fu*] (Precious Tools of Geomancy) *c.* AD. 1570.

Appearance of the magnetic compass in Europe and Islam

The first thing to be done is to note the exact dates at which knowledge of the magnetic compass first appears among Europeans and Arabs. As far as Europe was concerned, we find that the property of the lodestone to point in a given direction was unknown to Adelard of Bath in AD 1117, but was mentioned by Alexander Neckham in 1190. Many others then wrote about it, most notably Petrus Peregrinus (Peter the Wayfarer, or Peter de Maricourt) in 1269 in a text that is one of the finest contributions to physics of the whole medieval period. There have, of course, been claims for earlier European references, but all have been discredited, as too have suggestions that the Norsemen or Vikings had the magnetic compass earlier than 1190, though they do seem to have known of it by 1225. The suggestion has also been made that medieval European churches were orientated by means of the compass, but this is highly controversial and the fact has not yet been established.

The earliest Arabic references are all somewhat later than the European ones. The first mention is of sailors finding their way by means of a fish-shaped piece of iron rubbed with a magnet; it appears in a collection of anecdotes in Persian and compiled about 1232 by Muhammad ál-'Awfi. Half a century later, in a treatise on precious stones, Bailak al-Qabajaqi described how in 1242 he had witnessed the use of a floating compass needle, and added that the captains who sailed the Indian seas employed a floating iron leaf shaped like a fish. No other books of the thirteenth century discuss it, however; even the encyclopaedia by the geographer al-Qazwini is silent. There is no earlier mention of the compass, neither by tenth-century astronomers or geographers, though there is a treatise on love written early in the next century by Ibn Hazm that has been thought to refer to magnetism. Nevertheless this speaks only of magnetic attraction, not of the polarity of the compass. It may be noted, incidentally, that no Indian reference of any importance has been discovered.

Development of the magnetic compass in China

What then happened in China? To discover this it will be most convenient to take the basic text of the astronomer, engineer and high official, Shen Kua [Shen Gua], and then work back from it. This important passage appeared in the *Mêng Chhi Pi Than* [*Meng Qi Bi Tan*] (Dream Pool Essays) written about AD 1088, i.e. a little over a century before the earliest European mention of the magnetic compass. It runs as follows:

> Magicians rub the point of a needle with the lodestone; then it is able to point to the south. But it always inclines slightly to the east, and does not point directly to the south. (It may be made to) float on the surface of water, but then it is rather unsteady. It may be balanced on the finger-nail, or on the rim of a cup, where it can be

made to turn more easily, but these supports being hard and smooth, it is liable to fall off. It is best to suspend it by a single cocoon fibre of new silk attached to the centre of the needle by a piece of wax the size of a mustard-seed – then, hanging in a windless place, it will always point to the south.

Among such needles there are some which, after being rubbed, point to the north. I have needles of both kinds by me.

Moreover, the same book contains another, less well-known, passage:

When the point of a needle is rubbed with the lodestone, the sharp end always points south, but some needles point to the north. I suppose the nature of the stones are not all alike. Just so, at the summer solstice the deer shed their horns, and at the winter solstice the elks do. Since the south and the north are two opposites, there must be a fundamental difference between them.

This has not yet been investigated deeply enough.

Here, then, we have not only undeniably the earliest clear description of the magnetic needle compass in any language, but also a definite statement about magnetic declination. It greatly antedates the traditional discovery of magnetic declination by Columbus in 1492. The two kinds of needles mentioned by Shen Kua may of course have been magnetised at different poles of the lodestone, but there may also have been another origin for this traditional idea as we shall see (p. 21) when we discuss the divining board.

The modern scholar Wang Chen-To [Wang Zhen-Duo] has pointed out that some of Shen Kua's [Shen Gua's] experimental conditions indicate a considerable amount of careful investigation. For example, the use of new silk thread for suspension meant that a single fibre rather than one of twisted hempen yarn had finally been chosen, and its newness implied an appreciation· that its elasticity should be evenly distributed.

We must now turn to earlier texts though, before doing so, it is necessary to refer to one just a little later than that of Shen Kua [Shen Gua], but still many decades before the first European mention.

Sung compasses, wet and dry

In the *Pên Tshao Yen I* [*Ben Cao Yan Yi*] (The Meaning of the Pharmacopoeia Elucidated), a work which dates from 1116, there is a passage that seems on first sight to be but a repetition of what Shen Kua [Shen Gua] had said thirty years earlier, yet we find two things have been added. These read as follows:

Again if one pierces a small piece of wick (pith or rush) transversely with this needle (i.e. the magnetic needle), and floats it on water, it will also point to the south, but it will always incline (to the east) towards the compass-point Ping [Bing] (i.e. S15 °E). This is because Ping belongs to the principle of Fire, and the points Kêng [Geng] and Hsin [Xin] (in the west) which belong to Metal (the needle being of metal), are controlled by it.

Thus its (declination) is quite in accord with the mutual influences of things.

What the author Khou Tsung-Shih [Kou Zong-Shi] gives is, firstly, a description of the water-compass, so characteristic of all the oldest (but later) European accounts. Secondly, he not only gives a quite precise measure of the declination, but adds an attempted explanation for it.

Yet this is not the first description of the water-compass, for Wang Chen-To [Wang Zhen-Duo] has pointed out a passage in the great compendium of military technology, the *Wu Ching Tsung Yao* [*Wu Jing Zong Yao*] (Collection of the Most Important Military Techniques (compiled by Imperial Order)). Edited by Tsêng Kung-Liang [Zeng Gong-Liang] and finished in 1044, there is a passage that runs:

> When troops encountered gloomy weather or dark nights, and the directions of space could not be distinguished, they let an old horse go on before to lead them, or else they made use of the south-pointing carriage, or the south-pointing fish (chih nan yü [zhi nan yu]) to identify the directions. Now the carriage method has not been handed down, but in the fish method a thin leaf of iron is cut into the shape of a fish five centimetres long and a centimetre broad, having a pointed head and tail. This is then heated in a charcoal fire, and when it has become thoroughly red-hot, it is taken out by the head with iron tongs and placed so that its tail points due north. In this position it is quenched with water in a basin, so that its tail is submerged for several millimetres. It is then kept in a tightly closed box. To use it a small bowl filled with water is set up in a windless place, and the fish is laid as flat as possible upon the water-surface so that it floats, whereupon its head will point south.

This statement is full of interest. The fish-shaped iron leaf must have been slightly concave so that it would swim like a boat (see Fig. 167 showing Wang's reconstruction). What is more Dr Needham has noted that when a piece of iron is allowed to cool rapidly through the Curie point (600–700 °C) while oriented in the earth's magnetic field, the metal (especially if steel) will be magnetised. It seems extremely probable that this phenomenon, now called 'thermoremanence', was known to these Sung [Song] technicians; and though the magnetisation induced might be weak, it had the great advantage that no lodestone was required.

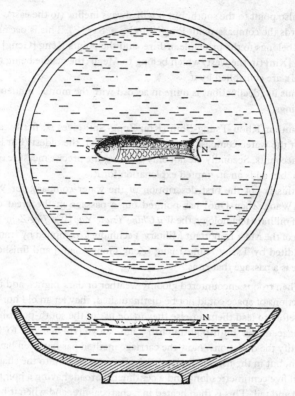

Fig. 167. Early forms of magnetic compass: the floating fish-shaped iron leaf described in the *Wu Ching Tsung Yao* (*Chhien Chi*) [*Wu Jing Zong Yao* (*Qian Ji*)] (Collection of the Most Important Military Techniques (compiled by Imperial Order)), AD 1044.

That heating destroys magnetisation must have long been known in China. It is mentioned in the *Wu Tsa Tsu* [*Wu Za Zu*] (Five Assorted Offering-Trays [miscellaneous memorabilia in five sections]) of 1600. But to return to the passage of the hollow fish, though the preface to the book containing it is dated 1044, it seems very likely that the description of the floating compass cannot be later than 1027. For it was in that year that the scholar, painter, technologist and engineer, Yen Su [Yan Su] succeeded in constructing a south-pointing carriage, and since our writer would have known of technological developments in the government workshops, he could hardly have said that the method of the south-pointing carriage had not been handed down. He may, of course, have been copying from some source earlier than 1027, but in any case the existence of the floating-compass seems clearly established for the first decades of the eleventh century, and in all probability would go back to the later part of the tenth. The fish form is particularly significant as we shall shortly see in connection with the first beginnings of the compass as a spoon. That the floating fish-shaped iron leaf spread outside China as a technique, we know from the description of Muhammad al-'Awfi just two hundred years later.

It seems certain, however, that while these things were going on, magnetite itself persisted in other forms of compass. These are described in the *Shih Lin Kuang Chi* [*Shi Lin Guang Ji*] (Guide through the Forest of Affairs) encyclopaedia, written by Chhen Yuan-Ching [Chen Yuan-Jing] who compiled it sometime between 1100 and 1250, and probably after the move of the capital to the south in 1135. In his section on the magical arts of the Taoist [Daoist] immortals he wrote

> They (the magicians) cut a piece of wood into the shape of a fish, as big as one's thumb, and make a hole in its belly, into which they neatly fit a piece of lodestone, filling up the cavity with wax. Into this wax a needle bent like a hook is fixed. Then when the fish is put in the water it will of its nature point to the south, and if it is moved with the finger it will return again to its original position.

Figure 168 shows Wang's reconstruction of this little object. The text then goes on to say:

> They also cut a piece of wood into the shape of a turtle, and arrange it in the same way as before, only that the needle is fixed at the tail end. A bamboo pin about as thick as the end of a chopstick is set up on a small board, and sustains the turtle by the concave undersurface of its body, where there is a small hole. Then when the turtle is rotated, it will always point to the north, which must be due to the needle at the tail.

This is of great interest because it indicates that the dry suspension from below was known during the Sung [Song] period. The reconstruction is given in Fig. 169. We shall later find (p. 33) that the floating compass remained more popular among the Chinese during the following centuries, however, and that when it was superseded by the dry suspension, this was connected with European maritime influence. Nevertheless the first dry suspension was Chinese. In both the types just described, we should note again the significance of fish and turtle shapes in relation to the spoon shape. What is particularly interesting is the association of a needle with the lodestone, not as a 'detached' pointer, but for making more precise, as it were, the orientation of the lodestone-carrying object.

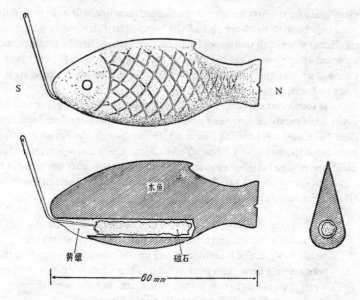

Fig. 168. Early forms of magnetic compass: the floating wooden fish containing lodestone and needle described in the *Shih Lin Kuang Chi* [*Shi Lin Guang Ji*] (Guide through the Forest of Affairs) (encyclopaedia). The lodestone and projecting needle pointer are fixed in place by wax.

References in the Thang [Tang] and earlier

Of reliably Thang [Tang] references to the magnetic needle we know of none, but there are many texts which were afterwards ascribed to this period. There is indeed a need for a systematic examination of Sung [Song] and Thang encyclopaedias and dictionaries for references to the lodestone and the magnetic needle. For instance, in a reference to the magnet in the *Shou Wên* [*Shou Wen*] (Analytical Dictionary of Characters) compiled in AD 121, the phrase *Shih ming kho i yin cheng* [*Shi ming ke yi yin zhen*] (The name of the stone which will '*yin*' the needle) has been shown to be a later addition. Yet it certainly occurs in the first printed edition which appeared in 986 in the early Sung [Song]. The use of the word '*yin*' is significant; its original meaning was to pull or draw a bow, hence to pull up a fish or to attract, or to light a fire. But it slowly came to mean lead, conduct, introduce or present, to derive, to propagate, to induce or to extend, even to prolong. There seems therefore no reason why it might not have meant, at some periods, the extension of directivity to the magnetised needle, as well as the attraction of the needle to the lodestone.

It has often been pointed out that if the magnetic compass had been used for navigation in or before the Thang [Tang], there should have been some

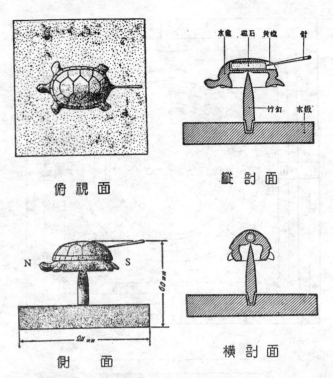

水盒　磁石　黃蠟　　　針

竹釘　水板

俯視面　　　　縱剖面

N　　　S

側　面　　　　橫剖面

Fig. 169. Early forms of magnetic compass: the dry-pivoted wooden turtle containing lodestone and needle described in the same source as Fig. 2. Above left, seen from above; below left, from the side; above right, section through length of body; below right, cross-section. The lodestone and projecting needle point are again fixed in place by wax, and the object is free to rotate upon a sharpened bamboo pin.

mention of it in the numerous accounts of the pilgrimages of Buddhist monks to India. No such mention has been found, but at that time the compass may well have been used purely for geomancy. However, it must be noted that in this literature there are references to the 28 *hsiu* [*xiu*] (the 'lunar mansions' of the sky) – see volume 2 of this abridgement – as azimuth points (i.e. directions around the horizon). These are still placed on the Chinese compass (Fig. 176) and indicate the region from S15 °E to S45 °E. This implies that the 'wind-rose' of the sea-captains of Thang times followed closely the principles embodied in the ancestor of the magnetic compass, the diviner's board. Indeed this is so old that in all probability the mariners of the Han had also been acquainted with it.

It this connection, it is worth noting that no mention of the mariner's

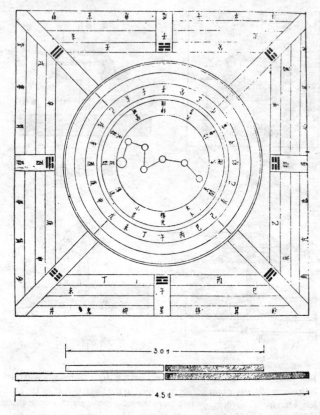

Fig. 170. The diviner's board (*shih* [*shi*]) of the Han period: diagrammatic reconstruction by Wang Chen-To [Wang Zhen-Do]. The apparatus was a double-decked cosmical diagram having a square earth-plate surmounted by a rotatable disc-shaped heaven-plate (see cross-section below the plan). Both plates were marked with compass-points, astronomical signs, etc. as well as the trigrams of the *Book of Changes* and certain technical terms used only in divination. The round heaven-plate carried a representation of the Plough (Dipper). The apparatus was of bronze or painted wood. North is at the top.

compass has so far been found in the accounts of China during the Thang [Tang] dynasty written by Arab travellers. Such negative evidence can carry but limited weight, for in all ages Chinese ships carried Taoist [Daoist] priests, and the use of the floating needle may well have been regarded as a numinous secret which would not readily be revealed to foreign merchants. It is interesting to remember that this negative evidence played a part in Jesuit controversies of the seventeenth and eighteenth centuries. One of their opponents published a translation of some ninth-century Arab voyages in China with the object of minimising as much as possible the philosophical and technological achievements of the Chinese. Yet Antoine Gaubil, the learned scholar and astronomer who was one of the Chinese mission, stated in the 1760s that he believed the compass had taken its definitive form about the time of the Emperor Hsien Tsung [Xian Zong] (about AD 800), and as we shall see, it may well be that he was not far from the truth.

We must now glance at two passages which have not previously been noted in the study of this subject. The first appears in the *Shu Shu Chi I [Shu Shu Ji Yi]* (Memoir on Some Traditions of Mathematical Art) attributed to Hsü Yo [Xu Yue] who worked about AD 150. Later research has shown that the book may in fact have been written by its sixth-century commentator Chen Luan, but even if this is so, the relevant text would be remarkable enough. It runs:

> In the Eight Kua method of calculation, a needle points at the eight directions. When a position is lacking, it points at Heaven.

Then comes Chen Luan's commentary:

> ... In this kind of computing, the digits are indicated by the pointing of a sharp end of a needle. The first digit occupies the *Li* position (i.e. pointing due south); second, or two, is at the *Khun [Kun]*, south-west; ..., and eight at *Sun* in the south-east. There is no place for the ninth digit, and so it is called 'perpendicular', (as if) the needle were pointing at Heaven.

This technique, which would seem to have been some simple sort of abacus-like device, arising out of the old diviner's board, is elsewhere attributed to or associated with Chao Ta [Zhao Da], a diviner of the third century AD. But the remarkable thing is that a needle is said to be used as a pointer, and the series starts from due south. It seems hard to believe that all this can have no connection with the magnetic compass, and the text must be at least AD 570 if not earlier.

The second is a strange passage, typical of its author, the alchemist Ko Hung [Ge Hong], who about AD 300 in the *Pao Phu Tzu [Bao Pu Zi]* (Book of the Preservation-of-Solidarity Master) wrote:

> Those who are prejudiced by affections do not bear criticism.
> Those who are envious of the beauty of others – their needle is not so bright and straight. Mugwort (i.e. ordinary) people are bewildered (by these affections) and have no 'south-pointer' to bring them back.

Here it is difficult to believe that the writer did not have some kind of magnetic compass in mind. It may, of course, be a subsequent addition, but if so it is not likely to be later than Chen Luan (*fl* AD 570).

The Han diviners and the lodestone spoon

Pushing back a few centuries earlier, we come to the *Lun Hêng* [*Lun Heng*] (Discourses Weighed in the Balance) of Wang Chhung [Wang Chong] (AD 83) and to the passage which Wang Chên-To [Wang Zhen-Duo] has so interpreted as to throw a flood of light on the origin of the compass. This runs as follows:

> As for the chhü-yi [qu-yi] (the 'indicator-plant' which was supposed to point accusingly at traitors to the Chou [Zhou] emperors) it is probable that there never was any such thing, and that it is just a fable. Or even if there was such a herb, it was probably only a fable that it had the faculty of pointing (at people). Or even if it could point, it was probably the nature of that herb to move when it felt the presence of men. Ancient people being rather simple-minded, probably thought that it was pointing when they really only saw it moving. And so they imagined that it pointed to deceitful persons.
>
> But when the south-pointing spoon is thrown upon the ground, it comes to rest pointing at the south.
>
> So also certain maggots which arise from fish or meat, placed on the ground, move northward. This is the nature of these maggots. If indeed the 'indicator-plant' moved or pointed, that also was its nature given to it by Heaven.

Here the author is contrasting a fable which he did not believe with actual events which he had seen with his own eyes. The description of the behaviour of the maggots may be regarded as the earliest mention of a tropism – an involuntary orientation by an organism – and what Wang Chhung had seen was probably some insect larvae of a species with a strong tendency to turn in response to the stimulus of light. The passage became garbled in some later editions.

The south-pointing spoon was nothing else than a piece of lodestone worked into that shape by jade-cutters, to imitate the shape of the constellation of the Plough (or Northern Dipper). It was used in conjunction with the diviner's board, which was itself composed of two boards or plates, the lower one being square (to symbolise the earth), and the upper one round (to symbolise heaven). The heaven-plate revolved on a central pivot and had the 24 compass points engraved upon it. At its centre it always carried a representation of the Plough (Dipper). The 'ground-plate' was marked on the edge with the 28 lunar mansions, and the 24 compass points repeated around its inner graduations. Moreover, it carried the eight chief trigram symbols from the *Book of Changes* (see volume 1 of this abridgement) arranged so that the *Chhien* [*Qian*] symbol occupied the north-west and the *Khun* [*Kun*] symbol the south-east. This is the same arrangement we met a moment ago (p. 17) in the curious computing machine mentioned in the *Shu Shu Chi I* [*Shu Shu Ji Yi*] mathematical text, but differs from that found on all later geomagnetic compasses, where *Chhien* [Qian] is south and *Khun* [Kun] the north. We shall consider this difference later. Figure 170 is a reconstruction of what the board looked like.

There is no dispute about the construction of the diviner's board. Fragments of actual examples have been found in Han tombs; these were made of flat pieces of wood cut to the correct shape and lacquered. One of the heaven-plates came from a tomb in which some objects were dated, giving the earliest date of burial as AD 69, and it is interesting that this was just contemporary with Wang Chhung's [Wang Zhong's] description of the lodestone spoon on the board. But another was some sixty years earlier, and among the items were spoons; though not of lodestone, their shape permitted easy rotation when balanced on their bowls.

Something is known of the wood from which the boards were made in later times – an eighth-century text, the *Thang Liu Tien* [*Tang Liu Dian*] (Institutions of the Thang [Tang] Dynasty), states that the heaven-plate was of maple, and the ground-plate of selected jujube wood. It may be significant that a particularly hard wood was chosen for the lower board, for that would be the surface on which the spoon would have to rotate if, as Wang Chen-To. [Wang Zhen-Do] suggests, the upper board was removed and the spoon substituted. Not all the ground-plates were made of wood, however; some bronze pieces of this kind may have been confounded with mirrors.

There is a close correspondence between the markings on the boards and the theories of the cosmos given in the *The Book of (the Prince of) Huai-Nan* (120 BC) previously mentioned (p. 3). In addition, an important section of the book is devoted to a survey of the characteristics and qualities of the seasons of the year as indicated by the handle of the Plough (Dipper) which, when observed at a given hour of the night, takes a year to shift round a complete circuit. It is clear, then, that that the handle of the Plough may be unhesitatingly denominated the most ancient of all pointer-readings, and in its transference to the heaven-plate of the diviner's board, we are witnessing the first step on the road to all dials and self-registering meters.

The fascinating problem of exactly why and how the heaven-plate was replaced by an actual model of the Plough (Dipper) we shall come to later (p. 24). Here, however, it should be noted that when Wang Chen-To [Wang Zhen-Do] began to investigate the nature of the metal spoon, various reasons led him to reject the possibility that it had been made of magnetised iron and resolve to explore the properties of a spoon made from lodestone. The only way he could find out whether such a spoon would have the strength to overcome friction to rotate on a ground-plate was by an experimental approach. One model that he constructed is shown in Fig. 171. He found that a tungsten

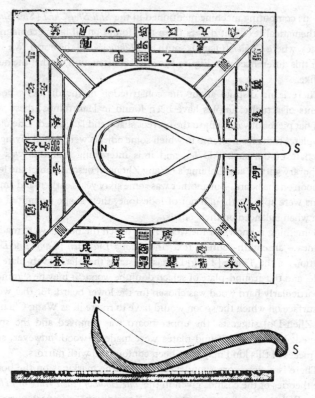

Fig. 171. Earth-plate and lodestone spoon: diagrammatic reconstruction by Wang Chen-To [Wang Zhen-Do], showing how the shape of the spoon permits rotation upon its bowl in response to magnetic pull.

steel spoon that had been magnetised as strongly as possible proved to be south-pointing, and would rotate on a ground-plate of bronze. It was not strong enough to do so on a hardwood ground-plate. He next tested a spoon made of lodestone from a source almost certainly available to the Han People; the results were equally conclusive.

In making their lodestone spoons, there was no need for the ancient makers to cut their block of magnetite from a vein of the material lying in the earth's magnetic field. This is because once a bar-shaped piece has been cut, the free poles at the ends of the bar will cause the magnetic polarity to lie along the bar's main axis. Such polarity would be intensified if the bar were heated and cooled, and this may well have occurred in the processes used by Han alchemists. On the other hand, since there was no way of testing polarity beforehand, the two ends of the bars may often have been confused, in which case some spoons would have had north-pointing handles instead of south-pointing ones. This might conceivably account for Shen Kua's [Shen Gua's] statement quoted earlier (p. 10) that there were two kinds of needles, some of which pointed north and others south. He may, of course, have been using two techniques which magnetised them differently, but on the other hand he may have been relying on an ancient tradition going back to the days of spoons, in which two kinds had been known.

It would seem almost impossible that any representation of the use of the lodestone spoon should have survived, but in a flat relief on a stone slab dated AD 114 which is in the van der Heydt Collection, now in the Rietberg Museum at Zürich (Fig. 172), one can see in the right-hand top corner a person apparently consulting what seems to be a large spoon of exactly the shape in question, not hidden by a bowl but resting almost fully in sight on the flat surface of a small square table. The rest of the relief, moreover, represents not a banquet, but some kind of ritual or entertainment accompanied by music and involving what may well be mechanical toys. Indeed, it looks like a field-day of the emperor An-Ti's [An-Di's] magicians and conjurers.

Literature on the diviner's board

We may now return to the consideration of further literary evidence. Two groups of references are to be considered, first those which speak of the diviner's board itself, and secondly those which speak of the south-pointer or south-controller.

As to the first, there is no controversy. The diviner's board was well known at all times from the Han onwards. Exactly how it was used has long been forgotten, and all that we can be sure of is that the 'heaven-plate' was turned on its axis in imitation of the imaginary passage of the handle of the Plough (Dipper) around the horizon according to the seasons. One of the earliest mentions must be in the *Tao Tê Ching* [*Dao De Jung*] (Canon of the Tao and its Virtue) which appeared in 300 BC. There is also a reference in the *Chou Li* [*Zhou Li*] (Record of the Institutions (lit. Rites) of (the) Chou (Dynasty)) which dates from about the second century BC and another, but later, Han work, the *Chhien Han Shu* [*Qian Han Shu*] to which we shall come later. Again the Han historian and astronomer Ssuma Chhien [Sima Qian] who worked about 100 BC wrote:

> Nowadays the diviners have a model of heaven and earth representing the four seasons and following benevolence and justice; in this way they group the bamboo stalks and decide upon what *kua* [*gua*] (trigrams and hexagrams from the *Book of Changes*) are involved. They turn the *shih* [*shi*] (diviner's board) to adjust the *kua* [*gua*]

Fig. 172. Han stone relief dated AD 114 and now in the Heydt Collection, Rietberg Museum, Zürich. Jugglers and acrobats are performing and music is being played; the figures beating upon the large drum in the centre may be mechanical jacks; indeed the whole structure may be some kind of roundabout or carousel automatically turning. The seated figures in the top row constitute perhaps an imperial audience. And still another wonder is presented to them in the extreme right top corner, where we see a large ladle of the same shape as the lodestone spoon, resting on the surface of a small square table (see small

drawing), and being attentively observed by a kneeling figure. It is possible that we have here, only a few decades after the words of the first century philosopher Wang Chhung [Wang Chong], a depiction of the earliest form of the magnetic compass.

to the right place. And then they are able to predict benefit or misfortune, success or failure, in human affairs.

One can see how natural it was that as the Plough (Dipper) revolved in the sky, the upper plate of the diviner's board should turn on the lower, and hence that the model Plough, or spoon, should also turn on it. Other mentions occur in the fourth, late fifth and sixth centuries, and Thang [Tang] encyclopaedias record at least three forms of the board. Bibliographies also contain references to texts concerned with its use; indeed, in AD 1150 one bibliographer was able to record no less than 22 books about it. The board was closely associated with alchemy – the ninth-century alchemist Tsung Hsiao-Tzu [Zong Xiao-Zi] used it to foretell the success of his chemical operations – and a study of the whole subject would be fruitful, even though it concerns a pseudo-science; after all, such subjects were the forerunners of modern scientific disciplines.

References to the 'south-pointer'

Of the texts which speak of the *ssu-nan* [*si-nan*] (south-pointer or south-controller), the oldest is in the *Kuei Ku Tzu* [*Gui Gu Zi*] (Book of the Devil Valley Master). This says:

> When the people of Chêng [Zheng] go out to collect jade, they carry a south-pointer with them so as not to lose their way. The assessment of abilities and probabilities is the south-pointer for human affairs.

Many editions of this book have slipped in the word *chhê* [*che*] (carriage) after *ssu-nan* [*si-nan*], but Wang Chen-To [Wang Zhen-Duo] regards this as a typical case where scribes or editors, unskilled in geomancy, had heard only of the carriage, and thought that every reference to south-pointing must refer to that. This did not always happen, though, for the original form of the text is sometimes preserved. That the writer of the quotation just given cannot have had a carriage in mind is shown by the use of the word 'carry'. There is, of course, little probability that the passage really goes back to the early fourth century BC, when its author Wang Hsü [Wang Xu] is supposed to have lived; it is almost certainly Han, but that does not lessen its value for the present argument.

More authentically pre-Han must be the remark in the *Han Fei Tzu* [*Han Fei Zi*] (The Book of Master Han Fei), which runs:

> Subjects encroach upon the ruler and infringe his prerogatives like creeping dunes and piled-up slopes. This makes the prince to forget his position and to confuse west and east until he really does not know where he stands. So the ancient kings set up a south-pointer, in order to distinguish between the directions of dawn and sunset.

Now distinguishing between these directions was what the vertical stake or gnomon, as well as the sun-shadow and the Pole Star are supposed to do according to the *Record of the Institutions of (the) Chou [Zhou] (Dynasty)* written in the early third century BC. If such astronomical methods had been in Han Fei's mind, these would surely have been mentioned rather than a special 'south-pointer'. As Wang Chen-To [Wang Zhen-Duo] has pointed out, it is in fact rather remarkable that the south-pointer never occurs among the lists of astronomical instruments. It seems likely that this was because it took its origin among a group of technicians quite different from the astrologers, namely the geomancers. In any case the words of Han Fei, part of a rather reliable text, carry the reference to the south-pointer furthest back, into the time of the rise of the Chhin [Qin] just before the Han.

Other instances of what were probably insertions by editors of the word 'carriage' occur in the *Chih Lin Hsin Shu [Zhi Lin Xin Shu]* (New Book of Miscellaneous Records) of about AD 325, written by the astronomer Yü Hsi [Yu Xi], and also in the seventh-century biography of Jen Fang [Ren Fang]. In addition there are a great number of metaphorical references to the south-pointer. The implication, then, is that since texts of this type written between the early Han (third century BC) and the late Thang [Tang] (ninth century) never mention the carriage, they must be referring to something else. Clearly it was some device, some compass-like instrument, and if it was not the lodestone spoon on the diviner's board, what could it have been? And if the beginning of it was not in the time of Wang Chhung [Wang Chong] and Chang Hêng [Zhang Heng] (first century AD), may it not perhaps go back beyond Luan Ta [Luan Da] to the late fourth century BC and the great school of the naturalist philosopher Tsou Yen [Zou Yan]?

The 'Ladle of Majesty'

One of the most striking historical events connected with the diviner's board was the latter end of Wang Ming, first and only emperor of the Hsin [Xin] dynasty, between the Former and Later Han. Part of the account of the final capture of his palace in AD 23 by the Han people, in the course of which he was killed, runs thus:

> (Wang) Mang fled ... to the Hsüan [Xuan] Room ... dressed all in deep purple and wearing a silk belt with the imperial seals attached to it, (he) held in his hand the spoon-headed dagger of the Emperor Shun. An astrological official placed a diviner's board in front of him, adjusting it to correspond with the day and hour. And (the emperor) turned in his seat, following the handle of the ladle, and so sat. Then he said, 'Heaven has given the (imperial) virtue to me; how can the Han armies take it away?'

In this impressive scene, the chief question for us is what exactly Wang Mang was following as he turned in his seat to face the magically correct direction. The most likely suggestion seems to be that either the ladle was the image of the Plough (Dipper) engraved on the heaven-plate or its lodestone model, the handle of which indicated to him the true direction he must face to remind the universe of his imperial virtue, and so to defeat the rebels.

The great importance of objects shaped like spoons and ladles in Wang Mang's time is indicated by another passage relating how he organised the making of a "Ladle of Majesty". It is not clear whether there was only one of these, or five, one for each of the five Chinese 'elements'. The event occurred in AD 17.

> ... (Wang) Mang went in person to the place for the suburban sacrifice south of the capital to superintend the casting and making of the Ladle of Majesty (*Wei Tou* [*Wei Dou*]). It was prepared from minerals of five colours and from copper. It was in shape like the Northern Dipper (Plough), and measured 0.75 metres in length. (Wang) Mang intended (to use it) to conquer all rebel forces by means of conjurations and incantations.... On the day that the Ladle of Majesty was cast, the weather was exceedingly cold, so that some officials and horses from the government offices were frozen to death (while in attendance).

There can be no doubt that this magico-ritual object, which was to be solemnly carried in processions when the emperor went out, was a model representation of the Dipper (Plough), just as the 'south-pointer' was a practical lodestone model.

From the spoon to the needle

Those who are acquainted with the literature on the mariner's compass in Europe are well aware that one of its earliest names was 'calamita'. While some have suggested that this was derived from the Greek word for reed (*kalamos*), and referred to the small piece of reed by which the needle was assisted to float, the more generally accepted view has been that the word meant a small frog or tadpole. It is thus used, for example, by Pliny. Those who can read Chinese are therefore liable to receive a considerable shock when they find, in the *Ku Chin Chu* [*Gu Jin Zhu*] (Commentary on Things Old and New), a fourth-century AD dictionary by Tshui Pao [Cui Bao], the following remarks:

> *Hsia-ma tzu* [*Xia-ma zi*], the tadpole, is also called the 'mysterious needle', or the 'mysterious fish', and another name for it is the 'spoon-shaped beastie'. Its shape is round and it has a long tail.

The text seems genuine, but in any case an almost identical passage with all the essential words, appears in the *Chung Hua Ku Chin Chu* [*Zhong Hua Gu Jin Zhu*] (Commentary on Things Old and New in China) which was written by Ma Kao [Ma Gao] in the tenth century. On the conventional view this is quite bad enough. There seems no other way of explaining them than to suppose that at some time between the fourth and tenth centuries the south-pointing lodestone spoon was superseded by a south-pointing iron 'fish' or 'tadpole', or needle, which had been magnetised. Then by a natural association of ideas, the needle came to be called a 'frog' or 'tadpole', while the tadpole itself acquired the popular name of 'mysterious needle'. That the Chinese character for 'tadpole' contains the spoon-ladle radical (tou [dou]) as its phonetic component is a point not to be missed. Certainly tadpoles and Chinese spoons are strikingly similar in shape, but the association of tadpoles with needles becomes much more understandable when one has reason to think that iron needles derived from lodestone spoons.

Finally, we know that between AD 1040 and 1160, i.e. long before European references, the Chinese were using (*a*) floating wooden fish-shaped objects containing lodestone, (*b*) floating fish-shaped iron or steel magnets, and (*c*), pivoted turtle-shaped objects containing lodestone.

Closer approximation to the date of origin of one of the Chinese methods of mounting a magnetic needle may be obtained by a critical study of textual variations. For instance, when one looks at the citation of the passage in the *Thai-Phing Yü Lan* [*Tai-Ping Yu Lan*] encyclopaedia (tenth century) one is astonished to find that the editors have written 'suspended needle'. Whether this change was made because the older term was not understood, or simply inadvertently, it betrays to us the fact that a full century before Shen Kua [Shen Gua], the system which he described of hanging the needle (presumably on a thread of silk) was known and used. The term employed in the encyclopaedia is *hsüan* [*xuan*], the same as that in the language of Shen Kua himself.

Certain other facts are probably also relevant. For instance the *Memoir on some Traditions of the Mathematical Art*, dating from the second to sixth centuries AD, and already referred to as important in the history of the compass, has a nearby passage referring to a method of computing by the use of the character *liao*. This is like a tadpole in shape. Secondly, in the Thang [Tang] (AD 766 to 779) Wei Chao [Wei Zhao] wrote an essay on scoops made from gourds and calabashes; in this he said that if one spins such a spoon-shaped object round like a top, it reminds one of the 'south-pointer'. The lodestone spoon was evidently still known or remembered at that date. Then it should also be recollected that the earliest Arabic references speak of the floating magnet as a 'fish'. Lastly, ancient head-and-tail ideas originating from the lodestone spoons were still being applied to the newer knowledge of magnetic polarity even as late as the eighteenth century.

Again, the better one gets to know medieval Chinese alchemical literature, the more valuable are the hints about the magnet which it is found to contain. A text deriving from between the late third to fifth centuries AD refers to 'needle-attraction on a fixed platform' as a name for magnetite, and this is interesting because the 'fixed platform' suggests that the writer had in mind the diviner's board, and even if this is fifth century it should be noted that this was a crucial time in the development of the compass-needle from the lodestone spoon. But there are many other references as Table 42 indicates.

Broadly speaking, alchemical references to iron are more numerous in the earlier periods, while after the Liang (AD 557 onwards) the lodestone is said to attract (or orient) needles rather than simply iron. Such a result is presented, of course, with all reserve, since copyists' confusions between 'iron' and 'needle' would be quite easy. If such mistakes be admitted, however, in the first two texts quoted in the table under the heading 'Needle', then evidence still suggests the first use of magnetised needles about the fourth century AD. This is a date that agrees with the bulk of other evidence so far assembled. It will, of course, also be the limiting factor for the discovery of magnetic declination, which could not be revealed by the lodestone spoon with its blunt end and clumsy frictional drag.

What first suggested a needle floating on water is an interesting question. It seems that there was an ancient method of divination which consisted in watching the shadow of a floating needle on the bottom of a bowl of water; indeed, this has been seen in recent times to be still in use at seasonal festivals among girls and young women in South China. There is good reason to place the technique at least as early as the Han because there is a reference to it in a Taoist alchemical book, the *Huai Nan Wan Pi Shu* [*Huai Nan Wan Bi Shu*] (The Ten Thousand Infallible Arts of (the Prince of) Huai-Nan), dating from the second century BC. Here it is recommended that to make the needle float it should be greased, as with sweat or the natural oil of the hair. This accompanies an entry about the lodestone.

THE USE OF THE COMPASS IN NAVIGATION

Here it will be best to proceed as we did with the magnetic compass, and allow the story to develop round a single text. This is the *Phing-Chou Kho Than* [*Ping-Zhou Ke Tan*] (Pingchow Table-Talk). Though written between AD 1111 and 1117, it referred to events from 1086 onwards, so it is approximately contemporary with the words of Shen Kua [Shen Gua] in his *Dream Pool Essays* already mentioned (p. 9). Yet there is no doubt that the author Chu Yü [Zhu Yu] knew what he was talking about for his father had been first a high official of the Port of Canton and then Governor. The passage runs:

Table 42. *Azimuth compass-points*

Modern compass-points °		Chinese names[a]		
0	N		Tzu [Zi]	
15		N15°E	Kuei [Gui]	
22·5	NNE			Tzu-Chhou [Zi-Chou]
30		N30°E	Chhou [Chou]	
45	NE		KÊN [GEN]	
60		N60°E	Yin	
67·5	ENE			Yin-Mao
75		N75°E	Chia [Jia]	
90	E		Mao	
105		S75°E	I [Yi]	
112·5	ESE			Mao-Chhen [Mao-Chen]
120		S60°E	Chhen [Chen]	
135	SE		SUN	
150		S30°E	Ssu [Si]	
157·5	SSE			Ssu-Wu [Si-Wu]
165		S15°E	Ping [Bing]	
180	S		Wu	
195		S15°W	Ting [Ding]	
202·5	SSW			Wu-Wei
210		S30°W	Wei	
225	SW		KHUN [KUN]	
240		S60°W	Shen	
247·5	WSW			Shen-Yu
255		S75°W	Kêng [Geng]	
270	W		Yu	
285		N75°W	Hsin [Xin]	
292·5	WNW			Yu-Hsü [Yu-Xu]
300		N60°W	Hsü [Xu]	
315	NW		CHHIEN [QIAN]	
330		N30°W	Hai	
337·5	NNW			Hai-Tzu [Hai-Zi]
345		N15°W	Jen [Ren]	

Kua are shown in capitals; denary characters in italics; duodenary in roman.

[a] The double terms shown in the third column of Chinese names may be found in geographical works, but the navigators hardly used them. For SSW their instruction was 'sail between Ting and Wei', not 'towards Wu-Wei'.

According to the government regulations concerning sea-going ships, the larger ones carry several hundred men, and the smaller ones may have more than a hundred men on board. One of the most important merchants is chosen to be Leader, another is Deputy Leader, and a third is Business Manager. The Superintendent of Merchant Shipping gives them an officially sealed red certificate permitting them to use the light bamboo for punishing their company when at sea. Should anyone die at sea, his property becomes forfeit to the government.... The ship's pilots are acquainted with the configuration of the coasts; at night they steer by the stars, and in the daytime by the sun. In dark weather they look at the south-pointing needle....

Here then is a very clear statement on the use of the mariner's compass just about a century before its first mention in Europe. But there has been a persistent theory that the account refers to foreign (Arab) ships trading at Canton, and that it was therefore Arabs who first saw the possibilities of the Chinese geomagnetic compass. This is quite erroneous; it originated because of a mistranslation whereby the words *chia ling* [*jia ling*] were thought to refer to the name of some foreign people, the 'Kling' (i.e. the Tamils in Singapore), whereas they are in fact a technical term for 'government regulations'. One has only to read the whole passage to see that the 'Kling' would not fit, for foreign merchants would not have required authorisation from the local Chinese authorities for disciplinary powers, nor would the property of a foreign merchant dying at sea have been forfeit to the Crown. Moreover, any assertion that the Chinese were not voyaging far afield collapses when adjacent passages are read, in which it says that repairs were often carried out at places in Sumatra.

There is another twelfth-century text which again refers to the use of the compass in poor weather and at night, and one of about 1123 by Hsü Ching [Xu Jing] concerning a Chinese diplomatic mission to Korea. The last reads:

... During the night it is often not possible to stop (because of wind or current drift), so the pilot has to steer by the stars and the Plough (Dipper). If the night is overcast then he uses the south-pointing floating needle to determine south and north. When night came on we lit signal-fires (to transmit compass-readings?), and all the eight ships of our convoy responded.

These words demonstrate once again that it was Chinese ships, and not those of other peoples, which carried the mariner's compass during the century before the first knowledge of it in the West.

Pausing for a moment to look, we see that the texts contain hints in abundance which, taken together, build up a picture of the development of the magnetic compass that may be indistinct but is certainly unmistakable. Moreover a study of the terms employed in the descriptions of magnetic attraction has given us strong indication that magnetised needles were coming in from the fourth to the sixth centuries AD. Later, as we shall see, there are texts of the ninth and tenth centuries which not only speak of needles but give values for the declination. Thus the picture unfolds.

On the other hand, so far as sailors were concerned, there is the negative evidence. A mention of steering by the stars occurs in the second century BC and later in accounts of travels by the fifth-century monk Fa-Hsien [Fa-Xian]. None of the Buddhist pilgrim voyages show anything else, and in the ninth-century navigational difficulties of a Japanese monk sailing from his country to Korea and China indicates that if the compass was at that time applied to maritime use, it must have been in possession of very few pilots.

It seems probable therefore that a rather long delay occurred between the first use of a magnetised needle in geomancy and its adoption by sailors for navigation. If it is correct, as Dr Needham believes, that the transfer of polarity from lodestone to needle was first made use of about the fifth century AD, it may well be that it was not applied to navigation before the tenth. The most probable period would be between 850 and 1050. Making a guess in the light of the evidence, one might say that the magnetised needle was probably used·for geomancy on an increasingly widespread scale from the late sixth to mid tenth centuries, not finding application at sea until the beginning of the Sung [Song] (960).

Before coming to a few interesting records of the Chinese use of the compass after AD 120, it is worth noting that the use of the compass in navigation depended to some extent on the methods used for the production of steel. Soft iron does not retain its magnetism for long, or show it strongly; for extended voyages magnetised needles of good steel would have been desirable. It seems probable that Chinese narratives of deep-water sailing which we have from the thirteenth century, such as an embassy to Cambodia, would not have been possible without steel needles. Now the history of steel is complicated enough, and the first use of it for needles still more obscure. The two most important ancient sources of for the material were Asia Minor and Hyderabad in India, and in a later volume of this abridgement where Chinese metallurgy is discussed, evidence will be brought forward to show that it was being exported from India to China at least as early as the fifth century AD. However, it will become clear that good steel was also being manufactured in China by remarkably modern methods at least from that time onwards. So it may well be that good steel needles were available to the Chinese several centuries before Europe had them. Figure 173 shows a scene of wire-drawing and needle-making from a later Chinese text.

Now for the later Chinese texts. Most famous is the *Chu Fan Chih* [*Zhu Fan Zhi*] (Records of Foreign Peoples) written by the Sung geographer Chao Ju-Kuan [Zhao Ru-Gua] in 1225. In this work he says:

> To the east (of Hainan Island) are the 'Thousand-Li Sand-Banks'
> and the 'Myriad-Li rocks', and (beyond them) is the boundless ocean,
> where the sea and sky blend their colours, and the passing ships sail

抽線琢鍼圖

Fig. 173. A scene of wire-drawing and needle-making using steel, from the *Thien Kung Khai Wu* [*Tian Gong Kai Wu*] (The Exploitation of the Works of Nature) of AD 1637.

only by means of the south-pointing needle. This has to be watched closely by day and night, for life or death depend on the slightest fraction of error.

Though the nationality of the ships is not mentioned, the whole context, which deals with Hainan, concerns an island which had been a Chinese province since Han times, and mentions junks coming from Chhüanchow (Quanzhou) in Fukien [Fujian] province and other Chinese ports. Half a century later a similar report is found in the writer Wu Tzu-Mu's [Wu Zi-Mu's] description of Hangchow (Hang-chou]:

> The water of the ocean is shallow near islands and reefs; if a reef is struck the ship is sure to be lost. It depends entirely upon the compass, and if there is a small error you will be buried in the belly of a shark.

Then in Yuan times (twelfth century) not only the compass but actual compass-bearings find their way into the literature:

> Embarking at Wênchow (Wen-chou in Chekiang province) and bearing SSW, one passes the ports of the coastal prefectures of Fukien and Kuangtung, thence ... one arrives at Champa (somewhere near Qi-Nhon in Vietnam) Thence with a good wind one can arrive in 15 days at Chen-phu [Zhen-pou] (somewhere near Cape St James), which is the frontier of Cambodia. Thence $S52\frac{1}{2}°W$ one crosses the Khun-Lun [Kun-Lun] Sea (north of Pulo Condor Island) and enters the river mouths.

Indeed, by the beginning of the Ming (mid fourteenth century), there must have been quite an abundant literature of sailing directions recording the compass-bearings, and still more is this the case for the numerous accounts of the famous voyages in the time of Chêng Ho [Zheng He] (1400 to 1431). Some idea of the skill of Chinese pilots at this time may be gained by the fact that in circumnavigating Malaya they laid their course through the present Singapore Main Strait, which was not discovered (or at least not used) by the Portugese till 1615 when they had been in these waters for over a century.

Oxford University possesses a manuscript of remarkable value in this connection, the *Shun Fêng Hsiang Sung* [*Shun Feng Xiang Song*] (Fair Winds for Escort) which dates from about 1430 when the great series of expeditions led by Chêng Ho [Zheng He] was ending. Besides a mass of general nautical information (tides, winds, stars, compass-points, etc.) it also contains a description of the use of the compass, as this short extract shows:

> If the wind rises from east, west, south or north, there may be a change of a whole compass-point. Those who observe this must

immediately take the proper steps. If it is a question of raising sail, this must be set in accordance with the compass bearing, changing according to necessity, sometimes more, sometimes less, and then when the wind becomes favourable, you work back to the original course.

It is interesting to note that the manuscript contains also a liturgy for use in the ship's chapel or before its shrine at the beginning of a voyage. In this the mariner's compass is prominent, and there is a litany in which the names of saints and sages contain a number of legendary and real Taoist [Daoist] geomancers. This is an added proof, if any be needed, that the mariner's compass was derived from the geomagnetic compass.

Actual examples of the Chinese mariner's compass are not very rare. Probably the oldest type includes those which are like flat bronze plates not more than 15 cm in diameter, having a bowl-shaped depression at the centre for water on which the needle is floated (Fig. 174).

The mariner's compass and the compass-card

Something remains to be said about the suspension used in the Chinese mariner's compass, and about the compass-card. If a magnetic needle is suspended within a circular box carrying the compass-points on its rim, the helmsman who wishes to direct his ship in a given direction has to keep the bow and stern in line with an invisible axis across the face of the compass. As time went on it was found convenient to fix to the pivoted magnet itself a light card on which the 'wind rose' with its compass-points was painted. The whole was then enclosed in a circular box which carried nothing but a line indicating the direction of the bow and stern of the ship. This allowed the helmsman to keep his ship on course with much greater accuracy.

However, the compass-card was not a Chinese invention; it originated in southern Italy, probably at Amalfi, shortly after AD 1300. Indeed, there is evidence to make it fairly certain that general use of the compass-card as well as the dry-pivoted compass, was transmitted from Europe in the latter part of the sixteenth century. This was probably done by Dutch or Portuguese ships first to Japan, after which it was gradually adopted by the Chinese later in the century.

On the other hand, it is clear that the first dry-pivoted compass, namely that described in the twelfth century *Guide through the Forest of Affairs* (p. 13 above) was Chinese. Presumably it travelled to Europe together with the floating needle or water-compass during the twelfth century. But while the floating needle was soon superseded in the West by the dry-pivot, in China it remained the most prevalent type both for geomancy and navigation until the middle of the sixteenth century (Fig. 175). The Italian compass-card then spread over East Asia, carrying with it the revived dry-pivot.

Fig. 174. Mariner's floating-compass of bronze, Ming period, diameter about 8.25 centimetres. South is at 'half-past ten o'clock', and a reference line on the bottom of the water-compartment seems to make allowance for a declination slightly west of north. The 24 compass-points are on the outer band, the 8 trigrams on the inner. From a paper by Wang Chen-To [Wang Zhen-Do] in 1951, vol. 5 (new series), p. 101, of the *Chinese Journal of Archaeology, Academia Sinica.*

Specimens of Chinese nautical and other water-compasses have been found and studied, and often comprise a separate 'teapot' which held the water to be poured into the central space. Moreover, Chinese ships carried lodestones for remagnetisation, and sometimes special kinds of water were prescribed, as in the *Fair Winds for Escort*, because there were superstitions about the precise manner in which the needle floated. Ceremonial libations were used at the time of preparing the compass.

Most landsmen would be inclined to put down the persistence of the floating-compass in Chinese ships to the celebrated conservatism of the

Fig. 175. Mariner's dry-pivoted compass, early Chhing [Qing] period, diameter about 8.9 centimetres, and now in the Museum of the History of Science, Florence. South is at the bottom. The 24 compass-points are in the outer band, the four cardinal points in the inner.

people. But with the usual irony of history, modern navigation came back to well-damped floating magnets in the end. In 1906 the Royal Navy replaced the compass-card by various forms of liquid compass in order to overcome the unendurable oscillations and vibrations of dry-pivoted magnets.

The direction-finder on the imperial lake

In what has gone before it has repeatedly been said that the south-pointing carriage and the magnetic compass had nothing whatever in common. While this is in general true, there are two places where the subjects touch.

One is in the story of the south-pointing boat which is mentioned first in an off-hand way in the *Sung Shu* [*Song Shu*] (History of the Sung Dynasty), where there is an important account of the history of the south-pointing carriage. It seems there was a lake in the palace gardens on which this boat used to sail. Of course this may have been no more than a popular legend based on the south-pointing carriage, but if the story had any basis in reality one might be inclined to see in it a very early application of the magnetic compass. For such rough experiments by palace magicians on the smooth waters of a lake, even the lodestone spoon would have done, and it is curious that the date (AD 500) corresponds rather well with that which has been suggested for the first beginnings of the floating magnetised needle.

The second place occurs in a biography of Tsu Chhung-Chih [Zu Chong-Zhi], the eminent fifth-century mathematician and engineer. It runs:

> ... When Wu of the Sung [Song] subdued Kuan-chung [Guan-zhong], he obtained the south-pointing carriage of Yao Hsing [Yao Xing], but it had only the shell, and no machinery inside. Whenever it moved, a man was stationed inside to turn it.

If this is to be taken as something more than deception, it implies that the man inside turned a visible pointer to coincide with south according to some information he had with him, and this could only have been a form of compass. Here again, if this was used only on ceremonial occasions under conditions where it moved on a smooth surface, then a lodestone spoon could have been used, though the date would allow of a floating 'fish' or even a needle being available.

MAGNETIC DECLINATION

In the quotations from Shen-Kua [Shen Gua] and Khou Tsung-Shih [Kou Zong-Shi], we found that not only the directive property of the magnet, but also its declination was described. Many other references in Chinese literature refer to this latter property, and these antedate the time at which even directivity was first known in the West. To appreciate this, we must examine the Chinese geomagnetic compass in its fully-developed form, because medieval knowledge of the declination is embedded in the traditional layout of this instrument.

Description of the geomagnetic compass

When first looking at the Chinese compass in its most elaborate form, one is bewildered by its complexity, and many Chinese scholars as well as most Western ones have been only too willing to dismiss it as a monument of superstitious web-spinning. Yet it is absurd to do so, for although geomancy itself was always a pseudo-science, it was nevertheless the true mother of our knowledge of terrestrial magnetism, just as astrology was of astronomy and alchemy of chemistry.

The geomagnetic compass is composed of some 18–24 concentric rings surrounding the small needle balanced at the centre. Looking first at the first and fifth circles starting from the centre (Fig. 176), we find the eight trigrams (*kua* [*gua*]) in the first, and then four of them written out in the fifth. If one consults a table of the trigrams in the *Book of Changes* (see, for example, Table 10 in volume 1 of this abridgement) a discrepancy is found between them. In the first ring of the compass the first trigram, signifying bright male creativity, occupies the south, while the second trigram, signifying dark female receptivity, occupies the north. In the fifth ring, the word for the first trigram lies in the north-west, and the second lies south-west. The interest of this is that the north-west and south-west positions were those on the divination board or *shih* [*shi*], the ancestor of the compass. In Han times there were two arrangements of the compass points with respect to the trigrams, and in the third century AD, just about the time that the lodestone spoon was giving place to the needle, it was felt that the old *shih* system was unsatisfactory and should be changed.

The fifth circle contains 24 divisions, and as Table 42 shows, these divide the circle into segments of 15 degrees each. The whole cycle of 12 characters – the so-called 'earthly branches', the *chih* [*zhi*] (see volume 2 of this abridgement, p. 184) – is represented here, though the cycle of ten characters – the *kan* [*gan*] – are omitted, i.e. precisely those shown twice over on the diviner's board. This omission left four spaces, and these were filled by the four most important *kua* [*gua*]. The remaining four of the eight trigrams do not appear on the written circle of points around the horizon or azimuth.

Compasses used by navigators reduced all this to its simplest form, using only 24 points, or even reducing them to 12 or 8. Since in China the geomagnetic compass long preceded the navigational one, and since the role of the cyclic characters around the azimuth was so old established, the 'wind-rose' of Chinese mariners probably developed in rather a different way from that of the West.

The three circles of Master Chhiu [Qui], Master Yang and Master Lai

If we now compare the fifth circle with the twelfth and the seventeenth, we shall see that the two outer ones appear staggered. For instance, the south point is moved $7\frac{1}{2}°$ eastwards in the twelfth circle, and $7\frac{1}{2}°$ westwards in the seventeeth. This has been shown to be due to discovery of magnetic declination during Thang [Tang] and Sung [Song] times, traditionally made by Chhiu Yen-Han [Qui Yan-Han], Yang Yün-Sung [Yang Yun-Song], and Lai Wên-Chün [Lai Wen-Jun]; the geomagnetic compass thus preserves these old observations in fossil form. And while we know that the idea that declination

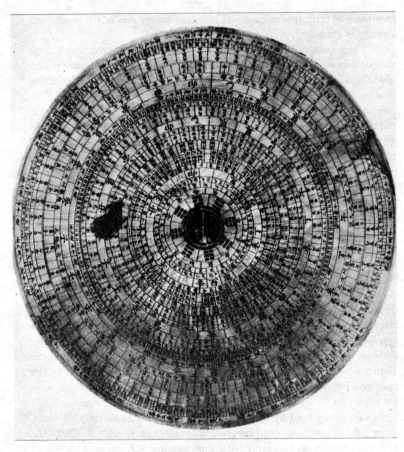

Fig. 176. The Chinese geomagnetic compass in full dress: an example
presented by the late Mrs John Couch Adams to the Cambridge Astronomical
Observatory, and now in the Whipple Museum, Cambridge. Diameter 32
centimetres. Details of the inscriptions are as follows:

*Translation of inscription painted by the maker on the back of the Chinese
geomantic compass presented by Mrs John Couch Adams, in the Wipple Museum
of the History of Science, Cambridge.* Made under the supervision of Wang
Yang-Chhi (courtesy name, Yuan-Shan) of the Yün-I Studio at Haiyang near
Hsin-an (in Kuangtung province). [Not dated, but of the Chhing, probably
late 18th century.]

IDENTIFICATION TABLE, FROM WITHIN OUTWARDS

      1   The Heaven Pool (a name derived from the medieval floating-compass,
          and from the ancient diviner's board), with the South-Pointing Needle.

1st circle   2   The Eight *Kua* [*Gua*] (Trigrams of the *I Ching* [*Yi Jing*] , the 'Book of
          Changes') arranged in the *Hsien Thien* [*Xian Tian*] (Prior to Heaven)
          order.

Caption for fig. 176 (*cont.*)

2nd circle	3	The Twelve Terrestrial *Chih* [*Zhi*] (cyclical characters).
3rd circle	4	Nine Stars (i.e. fate-categories, in 24 divisions) concerned with the Orientation of Dwellings.
4th circle	5	(24) Constellations controlling success in Civil-Service Careers (including the imperial examinations).
5th circle	6	Earth Plate (i.e. graduations controlling the Inner Region of the Compass Disc; with 20 cyclical characters and 4 *kua* as the standard Azimuthal Direction Signs) arranged in the *Chêng Chen* [*Zheng Zhen*] (lit. 'Correct Needle') positions (i.e. according to the astronomical north-south), with pure Yin and pure Yang (influences shown, by means of red and black colour. This arrangement is associated with the name of Chhiu Yen-Han [Qiu Yan-Han], 8th century).
6th circle	7	The (24) *Chieh Chhi* [*Jie Qi*] (fortnightly divisions) of the Four Seasons (the tropic year).
7th circle	8	Terrestrial Record (or Cycle) of the Mountain-Penetrating Tiger (60 combinations of the cyclical characters, i.e. the 12 *chih* [*zhi*] and the 10 *kan* [*gan*] , in 72 divisions – concerned with underground watercourses, veins, and foundations).
8th circle	9	The Nine Halls of the *Tun Chia* [*Dun Jia*] (method of divination; 72 divisions containing the numbers from 1 to 9, which also appear in the nine cells (Nine Halls) of the *Lo Shu* magic square [cf. Vol. 3, p. 57]. Twelve divisions are left empty as in the previous row. This method was popular in the 5th and 6th centuries).
9th circle	10	The *Fên Chin* [*Fen Jin*] of the Inner Part of the Compass (48 combinations of the cyclical characters, distributed among the Five Elements in 120 divisions).
10th circle	11	Orderly Arrangement, equally dividing the Dragon (Influence; 60 combinations of the cyclical characters in 60 divisions).
11th circle	12	The Five Elements (in 60 divisions) for the *Nei Yin* (method of divination; this dates from the 1st century BC).
12th circle	13	Man Plate (i.e. graduations controlling the Middle Region of the Compass Disc; with 20 cyclical characters and 4 *kua* as the standard Azimuthal Direction Signs, as in row no. 6) arranged in the *Chung Chen* [*Zhong Zhen*] (lit. 'Middle Needle') positions (i.e. all points shifted $7\frac{1}{2}°$ W of N as introduced in the 12th century, traditionally by) Lai Kung (Lai Wên-Chün) (to take account of the westward declination observed at that time).
13th circle	14	Mr Tshai's Earth-Penetrating Dragon (method; attributed to Tshai Shen-Yü [Cai Shen-Yu], 10th century; 60 combinations of the cyclical characters).
14th circle	15	Threes and Sevens riding the *Chhi* [*Qi*] (60 divisions containing 24 azimuthal compass-points corresponding with those in row no. 6; and 36 divisions containing odd numbers).
15th circle	16	The New Degrees adopted by the Present (i.e. the Chhing [Qing]) Dynasty (360 divisions, the numbering starting afresh at the beginning of each division of the following row, no. 17. The division of the circle into 360° was a Jesuit innovation, the Chinese having formerly divided it into $365\frac{1}{4}°$. At the same time the old Chinese division of the degree (*tu* [*du*]) into 100 minutes (*fên* [*fen*]), and the minute into 100 seconds (*miao*), was replaced, unfortunately, by the usual Western sexagesimal system.
16th circle	17	The (twenty-eight) Lunar Mansions (*Hsiu* [*Xiu*]; constellations dividing the Equator) with the number of degrees of equatorial extension occupied by each. (N.B. These two rows are placed with the

changed gradually with time only became conciously held during the Chhing [Qing] or Manchu dynasty in the eighteenth century, it is now clear that earlier, in the sixteenth century, the Chinese had reached the conclusion that the declination was different in various places. There was no clear statement, however, that this value changed with time until some time between 1736 and 1795.

Early observations of declination

We must now compare the traditional view of the discovery of declination with such texts as have come down to us. The learned nineteenth-century sinologist Wylie credited the Buddhist monk I-Hsing [Yi-Xing] with the first observation of it about AD 720, but the text to which he referred has not been found by later scholars. Yet there are.two old texts, both difficult to date, which also speak of a declination to the east. The first is from the *Kuan shih Ti Li Chih Mêng [Guan shi Di Li Zhi Meng]* (Master Kuan's Geomagnetic

Caption for fig. 176 (*cont.*)

		Middle Region of the Compass Disc because it agrees with the westward declination still observed in China in the 18th century, but as modern introductions they are separated from it by a double line.)
17th circle	18	Heaven Plate (i.e. graduations controlling the Outer Region of the Compass Disc; with 20 cyclical characters and 4 *kua* as the standard . Azimuthal Direction Signs, as in rows no. 6 and 13) arranged in the *Fêng Chen [Feng Zhen]* (lit. 'Seam Needle') positions (i.e. all points shifted $7\frac{1}{2}°$ E of N, as introduced in the 9th century, traditionally by) Yang Kung (Yang Yün-Sung) (to take account of the eastward declination observed at that time).
18th circle	19	The *Fên Yeh [Fen Ye]* (12 positions, i.e. the astrological controls of the different provinces, the names of which are given; a system popular in the 7th century; and the (12) *Tzhu [Zi]* positions (i.e. the 12 Jupiter-stations, their astronomical, not astrological, names being used for the years; they were also employed for the months of a single year and the double-hours of a single day. This circle is further evidence for a Thang date for this part of the compass).
19th circle	20	The *Fên Chin [Fen Jin]* of the Outer Part of the Compass (identical with row no. 10, but shifted eastwards. Another double line follows).
20 th circle	21	Celestial Record (or Cycle) of the Fully Coiled Dragon (60 combinations of the cyclical characters in 60 unequal divisions; corresponds to row no. 8 – concerned with hill ridges).
21st circle	22	The Five Elements (distributed among 61 unequal divisions) corresponding to the (Equatorial Extension) Degrees of the Constellations of the Celestial Sphere.
22nd circle	23	(Equatorial Extensions of the *Hsiu [Xiu]*, the Lunar Mansions, in) Degrees (Chinese), as determined in the Khai-Hsi [Kai-Xi] reign-period (1206).
23rd circle	24	Fortunate and Unfortunate Positions for the Siting of Tombs (indicated by red and black symbols; corresponding with row no. 4).
24th circle	25	(Equatorial Extensions of the) twenty-eight *Hsiu [Xiu]*, the Lunar Mansions, in (Chinese) Degrees for Divination (for divination because the $365\frac{1}{4}°$ system was that used in medieval times when the divination rules were fixed).

Instructor). Here we read:

> The lodestone follows a maternal principle. The needle is struck out from the iron (originally a stone) and the nature of mother and son is that each influences the other, and they-communicate together.... The nature of the needle is to return to its original completeness.... It responds to the *chhi* [*qi*] by orientation, being central to the earth and deviating in various directions. (To the south) it points to the Hsüan-Yuan [Xuan-Yuan] constellation (our Leo), hence to the *hsiu* [*xiu*] (lunar mansion) Hsing (part of our constellation Hydra)....

This is a perfectly definite statement that the declination was some 15 °E. The other early text also implies an eastern declination and comes from the *Chiu Thien Hsüan Nü Chhing Nang Hai Chio Ching* [*Jiu Tian Xuan Nü Qing Nang Hai Jue Jing*], a geomagnetic treatise with an almost untranslatable title that can perhaps best be rendered as 'The Nine-Heaven Mysterious-Girl Blue-Bag Sea Angle Manual'. The significant point here, of course, is the reference in the title to the sea. Obscure texts which are mentioned in the geomagnetic sections of encyclopaedias are difficult to date but the first is possibly late Thang [Tang] (ninth century) or about AD 900. The second is certainly Sung [Song] (i.e. second half of the tenth century), but if it was really Yang Yün-Sung [Yang Yun-Song] who introduced the first of the subsidiary staggered circles, then the passage it quotes would also be about 900.

An elaborate description of the geomagnetic compass appears in the introduction to the *Sea Angle Manual*, but that this should appear in the Sung [Song] merely fits in with the general pattern of development which we are gradually uncovering. Moreover, an additional argument which strengthens the relatively early datings of all these references to the magnetic needle, is the occurrence of the term 'suspended needle' in the 'tadpole' passage of the *Commentary on Things Old and New* quoted in the *Thai-Phing* [*Tai-Ping*] *reign-period Imperial Encyclopaedia* of 983, previously mentioned (p. 26).

What is probably the earliest datable reference to magnetic declination occurs as early as the first certain reference to the floating magnet. It appears in a poem by Wang Chi [Wang Ji], founder of the Fukien [Fujian] school of geomancers, whose main work was carried out between 1030 and 1050. It begins:

> Between (the *hsiu* [*xiu*]) Hsü [Xu] and Wei points clearly the needle's path
> (But) to the south there is (the *hsiu* [*xiu*]) Chang [Zhang], which rides upon all three;
> ... ...

The direction referred to in the first line is clearly the astronomical north–south line, but it can be seen by looking at a geomagnetic compass (e.g. the 24th circle, Fig. 176) that the southern *hsiu* [*xiu*] Chang [Zhang] is so broad that it can include all the three 'souths', i.e. those for the two declinations as well as the astronomical north–south line.

After 1050 all descriptions agree that the declination was to the west. However, one written about 1189 by Tsêng San-I [Zeng San-Yi] in his *Thung Hua Lu* [*Tong Hua Lu*] (Mutual Discussions) not only refers to the declination but proffers an explanation. He suggests that there must be some central place or meridian on the earth's surface where the declination would be found to be zero, a view that is more sophisticated than another which drew upon the Chinese theory of the five elements. Yet it can only have been an inspired guess, even though there is in fact a line of zero declination, which moves slowly across the earth's surface. All the same it is worth noting that Tsêng San-I was theorising about the declination before Europeans knew even of the polarity.

Knowledge of magnetic declination did not arrive in Europe until the middle of the fifteenth century. It was quite unknown in the thirteenth century to Pierre de Maricourt and his contemporaries, and the first evidence we have is that around 1450 German makers of portable sundials which embodied a compass for setting the noon line, were including a special mark on the dials to which the needle should point. The tradition that it was discovered by Columbus on his voyage of 1492 has thus been shown to be mistaken.

The reader may find it helpful to have a summary of the information contained in this section, and this is given in Table 43.

Traces in urban orientations

One last word may be devoted to the question of the orientation of city walls and the like. As early as 1763 the Jesuit Antoine Gaubil drew attention to the fact that the walls of Peking [Beijing], built about AD 1410, were $2\frac{1}{2}°$ out of the meridian to the west. Much later, about 1945, Mr Brian Harland who was working in Shantan [Shandan] city in Kansu [Ganzhou], one of the towns of the Old Silk Road, noticed that the street-plan seemed to show two separate alignments. These differed by about 11°, one due north – south and one due east. Many other Chinese city plans, those for Nanking [Nanjing], Chhêngtu [Chengdu], and Khaifêng [Kaifeng], for example, confirm such discrepancies. It seems quite possible that these differences are the results of medieval discussions by partisans of those geomancers who supported one or other of the methods of orientation which are to be found on the geomagnetic compass as a result of the effects of declination.

Table 43. Details on Chinese compass observations including magnetic declination, mainly before AD 1500

Date or approximate date	Author	Name of book	Probable place of observation	Lat. °	N '	Long. °	E '	Details	Declination °
c. 720	I-Hsing [Yi-Xing]	Unknown	Chhang-an [Changan] (Sian)	34	16	108	57	Between Hsü [Xu] and Wei 'to the right'	3–4 E
(c. 730)	Chhiu Yen-Han [Qiu Yan-Han]. The 24 azimuth points stabilised. Chêng Chen [Zheng Zhen]	This reference is very doubtful							
Mid 9th	Unknown	Kuan shih Ti Li Chih Mêng [Guan shi Di Li Zhi Meng] (Master Kuan's Geomantic Instructor)	Probably Sian	34	16	108	57	Ting-Kuei [Ding-Gui] axis	c. 15 E
c. 880	Yang Yün-Sung [Yang Yun-Song] adds new divisions each one 7½° to the left of Chhiu's, for eastern declination, and speaks of Fêng Chen [Feng Zhen] in his Chhing Nang Ao Chih [Qing Nang Ao Zhi] (Mysterious Principles of the Blue Bag (i.e. the Universe) [geomantic]))								
c. 900	Unknown	Chhing Thien Hsüan Nü Chhing Nang Hai Chio Ching [Jiu Tian Xuan Nü Qing Nang Hai Jue Jing] The Nine-Heaven Mysterious-Girl Blue-Bag Sea Angle Manual [geomantic]	Probably Sian	34	16	108	57	Speaks of Fêng Chen [Feng Zhen]	c. 7½ E
c. 1030	Wang Chi [Wang Ji]	In commentary on Kuan shih ... [Guan shi ....]	Probably Khaifêng [Kaifeng]	34	52	114	38		Slightly W
c. 1086	Shen Kua [Shen Gua]	Mêng Chhi Pi Than [Meng Qi Bi Tan] (Dream Pool Essays)	Khaifêng	34	52	114	38	'Slightly E' (of south)	5–10 W
1115	Khou Tsung-Shih [Kou Zong-Shi]	Pên Tshao Yen I [Ben Cao Yan Yi] (The Meaning of the Pharmacopoeia Elucidated)	Khaifêng	34	52	114	38	To Ping [Bing]	c. 15 W

Table 43. (cont.)

Date or approximate date	Author	Name of book	Probable place of observation	Lat. N ° '	Long. E ° '	Details	Declination °
c. 1150	Lai Wên-Chün [Lai Wen-Qun]	adds new divisions each one 7½° to the right of Chhiu's, for western declination, and introduces the term Chung Chen					
c. 1174	Tsêng San-I [Zeng San-Yi]	Thung Hua Lu [Tong Hua Lu] Mutual Discussions)	Hangchow [Hangzhou]	30 17	120 10	near Ping-Jen [Bing-Ren] axis	5–10 W
(1190 first knowledge of magnetic polarity in Europe)							
c. 1230	Chhu Hua-Ku [Chu Hua-Gu]	Chhu I Shuo Tsuan [Chu Yi Shuo Zuan] (Discussions on the Dispersal of Doubts)	Probably Hangchow	30 17	120 10	Ping-Wu [Bing-Wu] direction and Chung Chen [Zhong Zhen]	7½ W
c. 1280	ChhêngChhi [Cheng Qi]	San Liu Hsüan Tsa Chih [San Liu Xuan Za Zhi] (Three Willows Miscellany)	Probably Hangchow	30 17	120 10	Tzu-Wu and Ping-Jen [Zi-Wu and Bing-Ren] axes	7¼ W
(c. 1450 first knowledge of magnetic declination in Europe)							
c. 1580	Hsü Chih-Mo [Xu Zhi-Mo]	Lo Ching Chieh [Lo Jing Jie] (Analysis of the Magnetic Compass)	Probably Peking [Beijing]	39 54	116 28		c. 7½ W
c. 1625	J.A. Schall and Hsü Kuang-Chhi [Xu Guang-Qi]	The Magnetic Compass in China North China Herald Publishing House, 1859	Peking	39 54	116 28		5½–7½ W

Date	Author	Title	Place					
c. 1680	Mei Wên-Ting [Mei Wên-Ding]	Khuei Jih Chi Yao [Kui Ri Ji Yao] (Essentials of the Sundial)[a]	Nanking [Nanjing]	32	4	118	47	3 W
			Suchow. [Suzhou]	31	23	120	25	2¼ W
1690	de Fontaney	The Magnetic Compass in China	Canton	23	8	111	16	2½ W
1708	Régis & Jartoux	Observations Mathématiques, Astronomiques, Géographiques, Chronologiques et Physique tirées des Anciens Livres Chinois ou faites nouvellement aux Indes et à la Chine par les Pères de la Compagnie de Jésus, Paris, 1729[b]	Shanhaikuan [Shanhaiguan]	40	2	119	37	2 W
			Chiayükuan [Jiayuguan]	39	49	98	32	3 W
1817	Wylie		Canton	23	54	111	16	null
1829	Wylie		Peking	39	54	116	28	1½ W

[a] This reference is given by Wylie from a tractate which he did not quote in Chinese characters, giving only the title in his peculiar romanisation system. We conjecture it to be Khuei Jih Chi Yao. He says it was a 'small work on the sundial'. Unfortunately, it is not listed in the Wu-An Li Suan Shu Mu (Bibliography of Wu-An's Mathematical and Astronomical Writings) where, however, we find an Jih Kuei³ Pei Khao [Ri Gui Bei Kao] (A Study on the Construction of Sundials), a Khuei Jih Chhi [Ri Gui Qian] (Apparatus for Determining the Sun's Position), and a Khuei Jih Chhien Shuo [Kui Ri Qian Shuo] (Elementary Account of the Sun's Position). Perhaps Wylie was quoting one of these from memory. We have not been able to gain any further light from Li Nien's detailed biographical bibliography of Mei Wên-Ting, and must leave the matter to be cleared up by others to whom the works of the great Chinese seventeenth-century mathematician and astronomer are more accessible.

[b] It is not generally known that the Khang-Hsi emperor himself wrote on the declination of the compass. This was in his Khang-Hsi Chi Hsia Ko Wu Pien [Kang-Xi Ji Xia Ge Wu Bien] (Scientific Observations made in Leisure Hours), finished about 1710, and abstracted in French by Cibot (Mémoires concernant l'Histoire, les Sciences, les Arts, les Moeurs et les Usages, des Chinois, par les Missionaires de Pékin, Paris, 1779, 4. 452.) in 1779. The emperor said that the declination at Peking had been 3 °W in 1683, and had fallen to 2½°W at the time when he wrote; in some provinces an eastern declination was still observable. Further observations, also in the neighbourhood of Peking, varying up to 4½°W, were reported by Amiot in 1780 and 1782.

MAGNETIC VARIATION AND INCLINATION

In spite of the ancient legends about the attraction of ships with iron nails to certain islands which contained masses of lodestone, the discovery of real local magnetic variations was not a pre-Renaissance one in Europe. To William Barlowe (1597) we owe the realisation that ships' magnets were disturbed by iron in the vessel itself. Chinese sailors, however, seem to have been acquainted with local variation from the fifteenth century onwards. In his *Hsing Chha Shêng Lan* [*Xing Cha Sheng Lan*] (Triumphant Visions of the Starry Raft) of 1436 Fei Hsin [Fei Xin] remarked:

> There is a seamen's saying 'To the north we are afraid of the Seven Islands; to the south we fear the Khun-Lun [Kunlun] (Pulo Condor Island)'. At these places the needle may err, and if that happens, or the steering is inaccurate, both men and ships will be lost.

Then in a work as late as 1871, the *Tan-shui Thing Chih* [*Dan-shui Ting Zhi*] (Local History and Topography of Dan-shui (Formosa)), we hear of rocks· which bewilder the compass. How far back the knowledge of these effects goes we cannot now ascertain, but the description is closely similar to that of the classical experiments carried out by Portugese commander João de Castro on the island of Chaul off the west coast of India in 1539.

So far as it has been possible to ascertain, the vertical component of the earth's magnetic field – magnetic 'dip' or inclination – was never discovered in China. The first observation of it seems to have been due to Georg Hartmann in 1544.

THE MAGNET, DIVINATION AND CHESS

The whole story of magnetism in China so far has been extraordinary enough, but there is still a little more to be said. If we explore some dark by-paths in divination techniques from which games such as chess derive, we shall find indications that they alone contain the clue to the first use of a material earthly 'south-pointer'.

The essence of the problem is – how did the spoon get on to the diviner's board? Why should anyone have thought of making a model of the Plough (Dipper) at all, and of placing it on a board? That it should be made of lodestone, shaped from a bar of magnetite into the form of a dipper, is a thing less difficult to understand. But if it can be shown that there was an ancient form of divination which involved the use of 'pieces' resembling chess-men or something similar, and that these 'men' often represented celestial bodies, then the whole process of thought begins to reveal itself.

Due to the work of many scholars on the history of chess, the generally accepted view is that the war-game as we know it today was developed first during the seventh century AD in India. From there it radiated to Persia, to the Muslim world, and ultimately to Europe. But its antecedents have so far been very mysterious. For convenience we shall here define 'chess-men' as any collection, set, side or team, of small symbolic models which may represent anything, not only parts of an army, but animals or (significantly) celestial bodies and zodiacal 'houses'. Though China is the only civilisation where a close connection between the magnet and 'chess' can be shown to exist, the connection between chess and astronomical – astrological symbolism is widespread in all civilisations.

The fighting chess-men of.Luan Ta [Luan Da]

The first steps seem to have been taken in the second century BC. The text which first aroused Joseph Needham's interest and which led him to his conclusions, has come down to us in five versions, and is associated with the name of Luan Ta [Luan Da], one of the magicians of the Han emperor Wu Ti [Wu Di]. In the *Shih Chi* [*Shi Ji*] (Historical Records) of about 90 BC, we read:

> That spring (113 BC) the marquis of Lo-Chhêng [Lo-Cheng] presented a memorial recommending Luan Ta [Luan Da] to the emperor. ... ...
>
> The emperor was now regretting that he had put to death the Perfected-Learning General, and that the fullness of his arts had not been experienced, so he welcomed Luan Ta warmly. Luan Ta was tall and a brilliant talker, fertile in techniques, and daring in promises, never hesitating. ... ... At the end of the interview, the emperor asked Luan Ta to demonstrate one of his lesser arts by making chess-men fight automatically, and indeed they did mutually hit against each other.

The version in the *History of the Former Han Dynasty* written in AD 100 is identically worded. So far the lodestone has not been mentioned but it features prominently in the other three versions, all from the *Huai Nan Wan Pi Shu* [*Huai Nan Wan Bi Shu*] (The Ten Thousand Infallible Arts of (the Prince of) Huai-Nan) – a book of Taoist [Daoist] alchemical and technical recipes. If genuinely connected with the Prince, Liu An, it would be contemporary with the *Historical Records*. The excerpt preserved in the tenth century *Thai-Phing Yü Lan* [*Tai-Ping Yu Lan*] (Tai-Ping reign-period Imperial Encyclopaedia) is almost identical with that which the Thang [Tang] commentator Ssuma Chên [Sima Zhen] added to the text of the *Historical Records*. This runs:

> The Lodestone lifts (animates) chess-men.
> The blood of a cock is rubbed up with needle-iron (filings) and

pounded to mix. (Then when) lodestone chess-men are set up on the board, they will move of themselves and bump against each other.

Exactly how it was that the magnetised chess-men were moved is not clear, but the connection with needles is interesting, and suggests that the demonstration of polarity using needles may really have been older than we thought. But in any case the important thing is the association of the magnet with the men or pieces used in divinatory proto-chess.

Luan Ta's [Luan Da's] magnetised chess-men were not forgotten. The alchemist Ko Hung [Ge Hong] spoke of the use of them as an effective technique and speaks of using three sets of chess-men to foretell the success or failure of military enterprises. As he spent the latter part of his life in the Lo-fou Shan mountains north of Canton, we are not surprised to read in the eighteenth-century *Lo-Fou Shan Chih* [*Lo-Fou Shan Zhi*] (History and topography of the Lo-Fou Mountains), which is based on earlier accounts, that

> Under the Shih-Lou Fêng [Shi-Lou Feng] peak there is a stone as smooth as a mirror, and on it there used to be 18 chess-men, some black, some white. They moved to and fro and pushed each other about, yet if you tried to pick them up you could not do so. This was called 'spirit chess'.

Again, in 1126 Jen Kuang [Ren Guang] gave a term for chess-men which literally means 'white-jade magnetite objects'. And there are later references to wooden horses and paper men being made to dance by the use of magnets.

Some food for thought is given by western legends which parallel this material. In the Far West, St Augustine (AD 345–430) and an episcopal colleague, Serverus, both saw pieces of iron move about on a silver plate under the influence of a lodestone manipulated below. And even earlier, about AD 325 Julius Valerius making a Latin translation of a work falsely attributed to Greek historian Callisthenes, and which became one of the great sources of the medieval 'Alexander Romance', incorporated a very relevant passage. This was the legend that Alexander's father was the last of the Egyptian kings, Nectanebus the magician, who predicted the outcome of naval battles by the use of a basin of water on which floated little ships of wax containing model crews; these either began to move as if alive, or sank to the bottom, when he passed his ebony wand round the basin. Though the legend mentions neither lodestone nor iron, it would surely seem to have been invented by someone who had seen or practised such 'magic'. The story has a particular interest because it forms a 'pre-historic' background to the development of the floating magnetic needle which, as we have seen (p. 27), could have originated in China in the fourth century AD. Indeed, as Dr Needham remarks, one might venture to predict that somewhere in the vast mass of Chinese legendary material there will be found some parallel account of Luan Ta's [Luan Da's] chess-men turned marines and gone to sea. Nothing suggests that the European stories were not quite independent. But there is equally nothing to indicate that the knowledge behind them led to the invention of the floating-compass.

Chess and astronomical symbolism

Let us now approach the matter from another angle. It seems likely that the earliest European knowledge of chess was near the start of the tenth century AD, though there is no specific mention of it before the eleventh. The entry of the game to Europe was almost certainly by way of Spain from the Muslim world, where it had long been well-known. It also seems quite certain that the Arabs obtained it from India, where the earliest references to it occur early in the seventh century AD. There it had developed from an earlier game which had also used a checker-board and had probably been a race-game in which dice were used. Most of the authorities have considered that Chinese chess (using the word in its most precise sense) was derived from India, but their grounds for this are weak. The view rests, according to the chess historian, the late Harold Murray, 'on the identity of certain essential features of the two games, and partly upon what is known of the indebtedness of China to India in religion, culture, and above all, in games'. Yet board games are, of course, found everywhere, going back to at least the eleventh century BC in Egypt.

The oldest Chinese name for a chess-like game played on a board is *i* [*yi*] to which there are two references in the fourth century BC, but there is no sure information as to what it was, or how played. Yet from evidence of the first century AD, it may have been similar to or even identical with a war-game that used some 150 pieces a side; from the Three Kingdoms period (early third century AD) it was known as *wei chhi* [*wei qi*].

Chess proper – *hsiang chhi* [*xiang qi*] – became common during the Thang [Tang] some four hundred years later. The name has usually been interpreted as meaning 'elephant-chess', and there were generally four elephants among the pieces. But it can equally well be taken as meaning 'image' (*hsiang* [*xiang*]), 'model' or 'figure' chess, so-called in order to distinguish it from other earlier games in which all the pieces were identical. However, the earliest description of a game recognisably identical with chess proper occurs at the end of the eighth century AD in the *Yu Kuai Lu* [*You Guai Lu*] (Record of Things Dark and Strange). This contains the story of a man who dreamed he was present at a ceremonial battle, the moves being those of chess, and who afterwards found a chess set in an old tomb on the other side of the wall. This event is referred to in 762, but it has a background which, though known second or third hand to historians of chess, has hardly yet been appreciated at its true value.

The *Tan Chhien Tsung Lu* [*Dan Qian Zong Lu*] (Red Lead Record), compiled by Yang Shen shortly before 1554, summarises this as follows:

> Tradition handed down says that image-chess was invented by the Emperor Wu of the (Northern) Chou [Zhou] dynasty (AD 561–578). According to the *History of the (Northern) Zhou Dynasty*, it was in the fourth year of the Thien-Ho [Tian-He] reign-period (AD 569) that the emperor finished writing his *Image-Chess Manual*. He assembled all his officials in a palace hall and gave lectures to them about it.
> ... ...
> Then the story-tellers say that in the *Image-Chess Manual* (it was stated that) images of the sun, moon, stars, and constellations were used. From this it is to be supposed that the playing-board had ... (divinatory technical terms) marked on it. It was not the same as our modern chess, where chariots, horses, etc. are in play. Had it been like our modern chess, even ordinary people or children could have understood it without much difficulty. Yet it was necessary to have scholarly commentaries on it, and lectures to the hundred officials.

Confirmatory evidence of these traditions comes from another Ming scholar, Wang Shih-Chen [Wang Shi-Zhen]. And although the emperor's manual has disappeared, we are fortunate to have the preface which the divination-chess expert Wang Pao [Wang Bao] wrote for it. This runs:

> The first (great significance) of image-chess is astrological, for (among the pieces are represented) heaven, the sun, the moon and the stars. The second concerns the earth, for (among the pieces are represented) earth, water, fire, wood and metal.... The third concerns the Yin and the Yang ... The fourth concerns the seasons; ... The fifth concerns the following of permutations and combinations according to the changes of the heavenly bodies and the five elements. The sixth concerns the musical tones, following the dispersion of the *chhi* [*qi*]. The Tzu [Zi] position (among the compass-points) takes the cyclical sign *wei*, the Wu position takes *chhou* [*chou*], and so on. The seventh....

Here again is a definite statement that not only the heavenly bodies, but also the five Chinese elements were represented among the set of pieces. It looks also as if the positions of the pieces at the outset differed according to the position of the celestial bodies and the situation of the cyclical characters at the time when play was begun.

A parallel source is the essay on the *Image-Chess Game* by Yü Hsin [Yu Xin], a cavalry general of the sixth century AD. Using obscure language, he speaks of a board made round according to heaven, and another made square according to earth. This is valuable information because it links up the emperor's inventions with the diviner's board. He then speaks about the model pieces carrying ivory tablets like officials, and placed about according to stellar reference points. The boards had diagrams on them. Thus there seems abundant justification for the view that the more one investigates the origins of chess in Asia, the more intimate its connections with astronomy and astrology appear to be.

Before alluding to a few outstanding items in the mass of other evidence which associates chess with astrology and cosmic speculation, let us pause for a moment and try to reconstruct what the emperor Wu had in mind, though our ultimate object is to explain how a lodestone model of the Plough (Dipper) found itself on the diviner's board. But that occurred in the first century AD and by the sixth century, the time of emperor Wu, diviners had become far more sophisticated. Our immediate concern, therefore, is to see how a cosmic-astrological technique used for divination could have turned into a war-game for recreation. The answer is not far to seek; Wu's image-chess was nothing but a mimicry of the eternal contest between the two great forces in the universe, Yin and Yang. His desire was to determine the general balance between these in the existing cosmic situation, and if the model pieces were well chosen, their moves properly adjusted, and the board orientated and arranged according to concrete circumstances, the players, being themselves part of that situation, could not fail to proceed to a valid and informative decision. The idea of stars fighting against one another was quite old in Chinese astrology. Superstitious image-chess might be, yet it must have seemed at the time a brilliant device, evoking respect somewhat analogous to that given to today's computers.

For the Chinese mind, the good was essentially a perfect balance between Yin and Yang. However, it had always been realised, for example in medical circles when the cause of disease was discussed, that these two principles were not always in equilibrium. Image-chess was a way of detecting the extent of the unbalance at the time in question. It is easy to picture that, perhaps, the 28 lunar mansions were the pawns, while the two kings would be the sun and moon; the eight planets (eight because these included the old counter-Jupiter planet, and the two 'Indian' invisible planets [see p. 90 of volume 2 of this abridgement]) would be divided between the sides. Cannon and chariots (our knights and castles) may well have been comets. The remaining places may have been occupied by the five elements (perhaps represented on both sides), together with sundry bright stars such as Canopus or Algol. The 'river' dividing the Chinese chess-board across the middle still retains its original name of the Milky Way.

That this interpretation is on the right track is confirmed by no less an authority than Pan Ku [Ban Gu], the historian, who lived in the first century

AD and who understood the astrological significance, not of image-chess which had not been invented in his time, but of a game or technique which was probably *wei chhi* [*wei qi*]. In his essay on 'chess' he says:

> Northerners call *chhi* [*qi*] by the name *i* [*yi*]. It has a deep significance. The board has to be square, for it signifies the earth, and its right angles signify uprightness. The pieces (of the two sides) are yellow and black; this difference signifies the Yin and the Yang – scattered in groups all over the board, they represent the heavenly bodies.
>
> These significances being manifest it is up to the players to make the moves, and this is connected with kingship. Following what the rules permit, both opponents are subject to them – this is the rigour of the Tao [Dao].

He could hardly be more explicit.

Moreover, one can quote the converse. For the *History of the Chin* [*Jin*] *Dynasty* (seventh century AD) says:

> The heavens are round in shape like an open umbrella, while the earth is square like a chess-board.

To such an extent was the analogy fixed in the Chinese mind.

It only remains to be said that evidence supporting the division of the lunar mansions into two teams of fourteen each according to Yin and Yang may be found in Taoist [Daoist] books of the Sung [Song]. There are also a very large number of astrological 'pieces', some of which look like coins or medallions but also resemble the disc-like true chess-men described by Harold Murray. In addition, there exist numerous tokens representing the Dipper (significant because of the round heaven-plate of the diviner's board), and some show what may be other constellations (fig. 177). Again, there are larger discs with radiating arrangements reminiscent of the geomantic compass, and of the non-Chinese 'star-chess' which we shall come to shortly. Some of these give the azimuth points, others the eight trigrams or pictures of star-spirits. They may at some time have been used as temple tokens, but in any case the important thing to notice is that in China, and in China alone, the dominance of the Yin-Yang theory made it possible for a divination technique or 'pre-game' method to be devised. Only here could this contain at the same time both astrological and combat elements to enable it to be vulgarised into a purely military symbolism.

There is no need to commit ourselves to any definite conclusion as to when and where 'militarisation' of astrological image-chess took place; it may well have been in India in the following century. The appearance of 'elephants' may indeed have been a misunderstanding, since *hsiang* [*xiang*] can

Fig. 346   Fig. 347   Fig. 348

Fig. 177. Tokens resembling the pieces probably used in sixth century AD 'star-chess'. The one on the left depicts the Plough (Dipper). Above an inscription recalls the five male and two female spirits of its stars; the latter are represented to the left and right, the former are on the reverse. Below there is a sword, and then the tortoise and serpent, symbolic of the northern palace of the heavens. The three tokens on the right depict, left, the planet Mercury and its attendant spirits (both sides of the token are shown); right, the chih [zhi] cyclical character Wu, sign of the south among the compass-points and of the noon double-hour of the day. It is accompanied by its symbolical animal, the horse. All from the *Ku Chhuan Hui* [*Gu Chuan Hui*] (Treatise on (Chinese Numismatics) of 1859.

mean both 'image' (of a celestial body) and 'elephant'. It may even have been a substitute that sounded the same as another word meaning 'diviner'.

If the general conclusions about the origin of true chess are correct, we might expect to find widespread traces of astronomical symbolism clinging to it throughout later centuries. All historians of chess have agreed that this was in fact the case, though none of them has explained why.

There are many examples of this symbolism. In his *Muruj al-Dhabab ...* (The Meadows of Gold and Mines of Gems) of about AD 950, the famous Islamic geographer and traveller, al-Mas'udi, attributed the invention of chess to an Indian king, Balhit, saying:

> He also made of this game a kind of allegory of the celestial bodies, such as the seven planets and the twelve zodiacal signs, and allotted each piece to a star. The chess-board became a school of government; it was consulted in time of war, ...

What al-Mas'udi refers to was played on a square board, probably the direct descendant of the square ground-plate of the Chinese diviners. But what is extremely interesting is that there were several forms of chess played on disc-like boards with radial divisions, as if the round heaven-plate of the diviners also tenaciously lived on. Such boards were described by al-Mas'udi and seem to have been extemely popular in the Eastern Roman Empire; the game played on them was often known as Byzantine star-chess. In due course this found its way to western Europe. Astrological dicing boards were also related to it.

Divination by throwing

This provides the transition to an operation of the diviners not yet discussed, namely throwing objects on to a board, or tossing them on it or off. In astronomical image-chess or true chess, as in other board games, the pieces are moved on each side paralleling tactics and strategy. But in what are perhaps more primitive forms, the pieces were actually thrown on the board and conclusions drawn from where they came to rest. Thus the pieces approximated to dice, and no combat element was present. Hence the interest in references in Chinese literature to 'spirit-chess'. One we have is probably of Later Han or San Kuo [San Guo] date, and there is another from Sung [Song] times. Twelve symbolic pieces were used on a round wooden board, like the heaven-plate. A Sung encyclopaedia tells us that these pieces were thrown on the board for giving decisions as to good or evil fortune. But the author, Kao Chhêng [Gao Cheng] says that no one knows when the method was invented, and its origins seem lost in antiquity.

There was also a 'crossbow-bullet chess', which seems to have originated in the Han. It appears to have involved both throwing the pieces on to the board, and combat moves following this placing. There were twelve pieces – six for each player – and they seem to have symbolised the twelve celestial animal signs of the Chinese (see p. 191 of volume 2 of this abridgement). Later the pieces numbered twenty-four and in the Thang [Tang] Lu Tü [Lu Yu] wrote that 'the shape of the board is square below like the earth and round above like the heavens'. The pieces 'fly up when the board is quickly knocked, and scatter to different positions'.

Yet another game or divination-procedure which had a connection with astronomy was *liu-po* [*liu-bo*] ('the Six Learned Scholars') which was played with twelve 'chess' pieces on a board almost identical with the plate of the Han sundials. The moves of the pieces were determined by throwing six sticks, and they were divided into two 'sides', each piece being marked with one of the four animals symbolising the four directions of space. There seems to have been a central belt of water, like the Milky Way, in later systems. The relationship between these various games remains obscure.

Comparative physiology of games

We cannot here embark on a history of all Chinese games and divination techniques, but it is clear that from the earliest periods throwing things lent itself to divination as well as to games. One of the earliest was the 'pitch pot' game, where arrows were thrown into a pot.

Doubtless some day a social anthropologist will produce a fully developed and connected evolutionary story, probably biological in character, showing how they are all genetically connected. It would only need markings or numbers on the arrows to have an object which by compression would become a dice, and by extension or unfolding would give rise to dominoes on the one hand, and playing cards on the other. Cubical dice are ancient, examples having been found in Egypt and India, and it is generally assumed that they reached China from India; this we may accept. But it is now rather well established that dominoes and playing-cards were originally Chinese developments from dice. It seems that the transition from dice (leaf-dice, sheet-dice, etc.) to cards occurred at about the same time as the transition from manuscript rolls to paged books. Indeed, these cards, at their first appearance towards the end of the Thang [Tang] (tenth century), must have been among the earliest examples of block printing. After the beginning of the Sung [Song] about a century later, their evolution forked in two directions, one leading to playing-cards as we know them, the other to dominoes, from which again in its turn the famous game of 'Mah Jongg' derived.

Playing cards, certainly known in China before the end of the tenth century AD did not reach Europe until the fourteenth. There are no references to them in Arabic literature, but their dates would permit a direct transmission through merchant contacts with the Mongols more or less contemporary with Marco Polo. The case of dominoes is quite similar. It is said in the West that they were unknown in Europe until the eighteenth century, having been invented in Italy. But all Chinese manuals on dominoes, of which there are many, point back to the year AD 1120, when a set of thirty-two pieces was presented to the emperor. This was just before the move of the Sung [Song] capital to Hangchow [Hangzhou], a trading centre later to be visited by Marco Polo.

One important piece of evidence remains. At the turn of the last century, a study was made about chess and playing-cards and their related games, and of divination techniques in all civilisations. Much of this work was concerned with divination methods and gambling games widespread among the North American Indian tribes, where the pieces are tossed on to, or in, certain special baskets. What is significant is that many of the pieces were not simple counters, like the men used in draughts, nor even numbered dice, but were complex like chessmen. The Chippewa Indians, for example, had a set consisting of two human figures (rulers), one or two war-clubs, animals, and four plain counters. The pieces themselves might be marked with the four cosmic directions. This all encourages the view that the Chinese had sets of pieces symbolising the celestial bodies, and that divination took place by noting where they fell on a prepared board. In this way the model of the Plough (Dipper) placed on the divining board becomes quite comprehensible, and finds an appropriate context in divination methods in other parts of the world (Table 44).

GENERAL SUMMARY

Looking back over the arguments proposed, we see a long slow period of development in China, followed by sudden appearance and more rapid development in the West. We are compelled to recognise some transmission from east to west. But since the crucial couple of centuries before the thirteenth, when Alexander Neckham was the first in Europe to write on the magnetic compass in navigation, have yielded no trace or clue from intermediary regions such as the Arabic, Persian or Indian culture-areas, the possibility arises that this Chinese transmission occurred not in a maritime context at all. Rather it would seem to have come by some overland route through the hands of astronomers and surveyors who were primarily interested in establishing the meridian at various places. This was, of course, important not only in map-making but also for adjusting sun-dials, then the best European timekeepers. Thus Peter Peregrinus describes two instruments for this work, each with a built-in compass. It is certainly a striking fact that as late as the seventeenth century the needles used in the compasses of surveyors and astronomers were all made so as to indicate south (in contradistinction to the north-pointing sailor's needles), exactly as all the Chinese needles had done for perhaps as much as a thousand years previously.

We might then have to envisage an overland westward transmission of the 'astronomer's compass', followed by a western application of it by mariners independently paralleling the earlier application by the sea-captains of China. The level of culture in Central Asia in the two or three centuries preceding the Mongol invasions may have meant that what was transmitted might well have been thought of in a magical-technological rather than a scientific way by those who carried the information, but that presents no real difficulty.

In the developments described, we have strayed into fields which may not at first sight have any connection with physics as we understand it today. Yet we have in fact concerned ourselves with what is a fundamental problem, namely to discover the first origin of the ancestor of all dial and pointer-readings, the magnetic compass, as the following summary shows:

(1) The game of chess as we know it was associated throughout its development with astronomical symbolism. This was even more noticeable in games now long obsolete.

(2) The battle element in chess seems to have developed in China from a divination technique for determining the balance between Yin and Yang

57

Table 44. *Chart to show the genetic relationships of games and divination-techniques in relation to the development of the magnetic compass* Chinese examples underlined

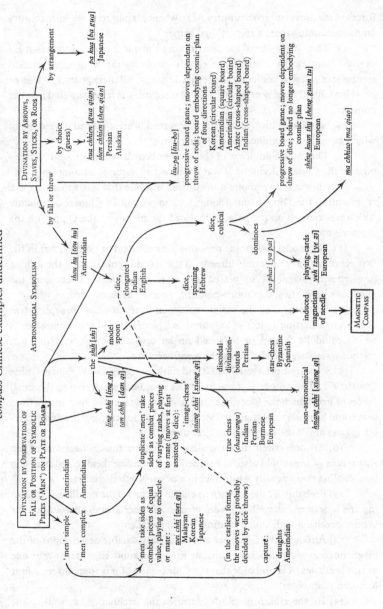

forces in the universe (sixth century AD), whence it passed to seventh-century India, generating there a recreational game.

(3) The 'image-chess' for determining Yin and Yang balance was itself derived from earlier divination methods involving throwing small models symbolic of celestial bodies on to prepared boards. All these go back to Han or pre-Han China (third century BC). Similar techniques have persisted in other cultures.

(4) Numbered dice, anciently widespread, were a related development which gave rise to dominoes and playing cards.

(5) The most significant of the ancient boards was the round heaven-plate of the Chinese diviners. It was the direct ancestor of all compass-dials.

(6) Among the symbolic models used was one that was spoon-shaped, representing the Dipper (our Plough), so important in Chinese astronomy which was centred on the pole. It replaced the picture of the Dipper on the diviner's board.

(7) The model spoon was probably first of wood or pottery, but in the first century AD (or possibly already in the second century BC), the unique properties of magnetite suggested its use to the Chinese. Because of the material's properties, the 'south-pointing' spoon was discovered.

(8) Later the way the magnetite spoon dragged on its baseplate was avoided by inserting a piece of lodestone in a piece of wood with pointed ends which could be floated or balanced on an upward projecting pin. Such methods were used as late as the thirteenth century.

(9) Some time between the first and sixth centuries it was found in China that the directive property of the lodestone could be transferred to the small pieces of iron which the lodestone attracted. Small pieces of such magnetised iron could be made to float upon water by suitable devices, and the first water-compasses are of the early eleventh century.

(10) At some date before the eleventh century the Chinese discovered that pieces of iron could also be magnetised by raising them to red heat and then cooling them rapidly while held in a north-south direction.

(11) Probably by the seventh or eighth centuries the needle was replacing the lodestone, as well as pieces of iron of other shapes, because of the greater precision with which readings could be taken.

(12) Although the first clear and accurately datable descriptions of the magnetic compass with needle, antedate European knowledge of it only by one or two centuries, it is probable that the Chinese use of it is some three or four hundred years older.

(13) By the eighth or ninth centuries the declination, as well as the polarity, of the magnet had been discovered. This antedates the European discovery of declination by some six hundred years. The Chinese were theorising about the declination before Europeans knew even of the polarity.

(14) The successive declinations, first eastern and then western, were embodied in the design of the Chinese geomantic compass as concentric circles, and these have persisted until our own time.

(15) The compass was undoubtedly employed in China for geomancy a long time before it was used to assist navigation.

(16) The first clear and accurately datable description of the use of the compass for navigation on Chinese ships antedates the knowledge of this technique in Europe by just under a century. However, there are indications that it was used for this purpose by the Chinese somewhat earlier.

(17) Chinese sailors remained faithful to the floating compass for many centuries. Although the dry pivoted compass had been described early in the twelfth century, it did not become common on Chinese vessels until re-introduced from the West four hundred years later by the Dutch and Portuguese by way of Japan. Associated with it was the compass-card or wind-rose attached to the magnet; this was probably an Italian invention made in the fourteenth century.

(18) The ancestor of all dial- and pointer-readings, the greatest single factor in voyages of discovery, and the oldest instrument of magnetic-electrical science, may thus perhaps be said to have begun as a 'proto-chessman' used in a method of divination.

(19) Magnetical science was an essential component of nascent modern science, but unlike geometry and planetary astronomy, its antecedents were not primarily Greek. All the preparation for Peter de Maricourt, and hence for the later ideas of Gilbert about the Earth as a magnet and of Kepler on a role for magnetism in astronomy, had been Chinese. In their turn, with their belief that gravity must be something like magnetic influence, Gilbert and Kepler played an important part in the preparation for Newton. The field of physics of still later times, firmly established by the mathematical equations relating electricity and magnetism which Clerk Maxwell derived in the nineteenth century, firmly established a more organic way of looking at the natural world than did the atomic materialism of the Greeks. This, too, can be traced back to the same root. Thus much is owing to the faithful and magnificent experimenters of medieval China.

2

Nautical technology

In the last years of the seventeenth century, the Jesuit missionary Louis Lecomte wrote:

> Navigation is another point that shows the address of the Chinese; we have not always seen in Europe such able and adventurous Sailors as we are at present; the Ancients were not so forward to venture themselves upon the Seas, where one must lose sight of Land for a long time together. The danger of being mistaken in their Calculation (for they had not then the use of the Compass) made all Pilots circumspect and wary.
>
> There are some who pretend that the Chinese, a long time before the Birth of Our Saviour Christ, had sail'd all the Seas of India, and discover'd the Cape of Good Hope; However that may be, it is most certain, that from all Antiquity they had always stout Ships; and albeit they have not perfected the Art of Navigation, no more than they have done the Sciences, yet did they understand much more of it than the Greeks and Romans; and at this Day they sail as securely as the Portuguese.

The remarkable justness of this appraisal will, it is hoped, become apparent towards the end this volume.

The history of Chinese nautical technology can mean little unless set forth in a comparative way, so that its special contributions can be seen, for it certainly had its own distinctive characteristics. Indeed Chinese shipping has generally been thought to stand very much on its own. Whether priorities in time also imply transmissions, even if only of ideas, to other peoples is an acute problem, and a final answer may prove impossible.

The Chinese texts about shipping speak for themselves, though only a few can be included here. The sources are many and various, for unlike some other technical subjects, such as agriculture or pharmacy, systematic nautical treatises did not arise in Chinese culture, or at least did not get into print. First there are ancient dictionaries and encyclopaedias; these are primary sources for sea terms, though later compilers of lexicons are also of use. Secondly, drawings and paintings abound, though these are strongly oriented towards warships rather than civilian traders, doubtless because the official bureaucracy were concerned with naval matters, whereas the others could look after themselves. Oldest in this line is the *Ching Tsung Yao* [*Jing Zong Yao*] (Collection of Military Techniques), an encyclopaedia by Tsêng Kung-Liang [Zeng Gong-Liang] which was published in AD 1044, though as we shall see, its descriptions go back to three hundred years earlier. It was truly a focal point, for from it stemmed the pictures not only of a series of later naval handbooks, but also those of a long succession of naval sections of encyclopaedias on the grand scale. These in their turn were gathered in by eighteenth-century Japanese books which combined them with very different illustrations of ships of the island culture, ships which embodied an altogether different tradition of naval architecture.

Of course, passages about shipping of one kind or another are liable to turn up almost anywhere in Chinese historical writing, whether in the dynastic histories themselves and the great compilations based on them, or in unofficial records and private memoirs. An oustanding example of such memoirs is the *Phing-Chou Kho Than* [*Ping Zhou Ke Tan*] (Pingchow Table-Talk) written in 1119 by a man whose father had been (as we have seen, p. 27) Port Superintendent and later Governor of Canton; this deals with the maritime life of the coast during the last decades of the eleventh century. Naturally enough, there is also much about shipping to be found in literature dealing with foreign countries, and perhaps the best of all classical depictions of a Chinese sea-going ship is contained in the eighteenth-century *Liu-Chhiu Kuo Chih Lüeh* [*Liu-Qiu Guo Zhi Lüe*] (Account of the Liu-Qiu Islands) by Chou Houang [Zhou Houang] and reproduced in Fig. 184.

For navigation and shipbuilding we are unusually dependent on manuscript material, and we have already seen (pp. 270–1 of volume 2 of this abridgement) how sailing directions and sea-charts were beginning to be preserved in the late Sung [Song], Yuan and Ming, after the use of the mariner's compass had become universal on Chinese ships. By the seventeenth century substantial books on navigation were printed both in China and Japan. For shipbuilding we are again beholden to a remarkable though very late manuscript (now in the famous Staatsbibliothek at Marburg in West Germany). This is the anonymous 'Fukien [Fujien] Shipbuilding Manual', which belongs to the end of the eighteenth century. As to Chinese literature itself, though this is not rich in works dealing specifically with the construction of ships, there is abundant material on shipping in China to be found in the works of sailors and scholars from our own western maritime civilisation.

So much for the texts. As we shall find, archaeological evidence comes to our aid, in the form of both pictures and models. This we shall mention as it arises.

FORM, STRUCTURE AND EVOLUTION OF SAILING CRAFT

What is the fundamental relationship of the Chinese junk to all other types of craft which men have used? This can best be understood by means of a summary such as that contained in the accompanying chart (Table 45) based on a survey by the naval historian J. Hornell.

As will be evident, the chart is arranged according to the various natural or artificial objects which ancient people must have seen floating on the water, and upon which they thought of launching forth themselves. At the outset, then, we must dispose briefly of a number of primitive types of craft which had no great future before them. Thus floating baskets gave rise to a number of small boats, some of which are still in use today. They could simply be caulked, either with bitumen if this was available, or otherwise with a kind of mud. When the basket was covered with skin or hide, the familiar coracle came into existence, a vessel that seems to have a Mesopotamian origin. Though no evidence of this simplest form remains in China proper, the 'skin boat' as the Chinese called it was the characteristic vessel of the rapidly flowing waters at the head of the great rivers which take their rise in Tibet. Apart from the Tibetan borders, they are now used only in Manchuria and Korea. Yet this was not always so. In the fourth century BC there was a reference to a boat which a man could come and carry away, and in the fourth century AD cowhide coracles were used in considerable numbers to mount feint attacks during military operations along the Yellow River. Three coracles are depicted on the walls of a Sui (late sixth century AD) cave at Chhien-fo-tung [Qian-fo-dong] and as late as the thirteenth century boats of this type were abundantly used by the Mongols in their conquests. But in the Chinese culture-area such craft never developed into the elongated skin-covered and decked boats of the Eskimos and northern Siberian peoples.

Other basket-shaped objects may also float. The pottery bowl, if large enough, may be used to carry a man, as in Bengal, but its place is taken in China by the wooden tub, the *hu chhuan* [*hu chuan*] or 'kettle-boat', and is much used for collecting aquatic food plants.

Buoyant objects, more or less spherical, formed the beginning of another line of descent. Swimmers supporting themselves on gourds or inflated skins are frequently to be seen on Assyrian reliefs of the ninth century BC, and the method survived for many hundreds of years in many parts of the world, especially for campaigning purposes, as in the Mongol armies, whence it entered Chinese military books from the eleventh century AD onwards. From the next century onwards, they were known in China as 'waist-boats'.

The transition to what might be called craft in the strict sense came when a number of buoyant objects were attached to a framework of wood to form a floating raft, which seems to have had its origin in the region of fast-flowing rivers in Central Asia. Yet even now the *phi fa tzu* [*pi fa zi*] (skin-raft), composed of thirteen goatskins, is still common in north-west China, on the Yellow River and its tributaries. One can still see them carried on men's backs. Closed pottery vessels as floats seem not now to be used in China, but they certainly were in antiquity, for the Han general Han Hsin [Han Xin] achieved a famous crossing of the Yellow River with his army by the aid of such rafts.

The real key point in the origin of most ships is thought by many to have been the observation of the single floating log, and its conversion to a boat by hollowing out to form a dugout canoe. It would then be a natural notion to increase the 'freeboard'–by adding first a piece of wood to form a small side wall, and later gradually building upwards with a succession of such 'strakes' to become the sides of a ship. Thus in most kinds of ship the ghost of the dugout canoes lives on in the shape of the keel which still contributes to the necessary lengthways strength of the vessel and, if projecting much below the hull, is important for the craft's sailing qualities. An early invention was to flare the sides of the dugout underbody, the flared sides being kept in place by inserting U-shaped frames. Finally the keel became purely a beam, and from the forward end there grew out of it the stempost, and at the rear the sternpost. Boat-building then diverged into two recognisably different traditions. In one the strakes overlapped one another, to give the 'clinker-built' boat; in the second tradition they were fixed edge to edge so as to form the comparatively smooth outer and inner surfaces of the 'carvel-built' vessel.

This classification is one that will cover all vessels, but it is far from being the whole story. Broadly speaking, the clinker-build is characteristic of northern Europe alone, while ships of other regions – the Mediterranean, the Persian-Arab culture area, India, Oceania, and China – are of carvel type. Further distinctions arise; the strakes may be sewn together with vegetable fibres, or they may be nailed or secured by wooden pins. In addition, the strakes may be fitted round the frames, or the frames inserted afterwards.

Compared with the distinction we now have to make, all these details are relatively unimportant. For it is clear that the ships of East Asia cannot be explained on the theory of a development from a simple floating hollow log. Bamboo is their ancestral material, not wood at all, for as we shall see, the Chinese hull (however the sides are built) is an elongated structure as full of bulkheads across it as the stem of the bamboo is of those partitions which the botanist calls 'septa'. It is this form of construction which sets the Chinese ship apart from the ships of the rest of the world.

It must have occurred very early to primitive men that instead of hollowing out a single log to make a canoe, they could bind a number of logs together and obtain a 'ship' of considerable size. Rafts made of bundles of

Table 45. *Chart of the development of boat construction* (based on studies by J. Hornell)

Floating objects

elongated objects

- **reeds**
  - reed rafts, 38, 69
  - ancient Egyptian boats, 48
  - modern tankwa, etc. (Upper Nile), 53
  - Titicaca balsa, 41

- **bamboos**
  - bamboo rafts, 69, 76, 78, 79 (with centre-board, 80, 82) ⊕
  - (CARVEL-BUILT) JUNK
    - ⊕ only, 85 ff.
    - sf/n
    - outboard poling gangways, 253, 254

- **log**
  - several
    - ⊕ log rafts, 61 ff.
    - sailing catamaran, 61 ff.
  - single
    - double canoe
    - dugout canoe, 189
      - added strakes abutted
      - added strakes overlapping
  - CARVEL-BUILT BOAT or SHIP, 189
    - Indian, 192, 236, 248, 249 — if/s if/n
    - Pers./Arab., 234, 235 — if/s
    - S. Eu. (Med.), 194 — pf/n if/n
    - balance-boards, 260
    - outriggers, single and double, 253, 254
  - CLINKER-BUILT BOAT or SHIP, 195
    - Oceanian, 207 — if/s if/n
    - N. Eu. 196, 199 — if/s if/n
    - cleats on strakes, bifid stem and stern, special bailers, etc.

fruit, gourd, inflated skin, or closed earthenware pot
- ⊕ swimming, 13

float raft, 20
- skins, 25 ⊕
- pots, 37 ⊕
- pontoons

half-fruit
- ⊕ tub or half-barrel, 108
- pottery bowl, 98

basket
- caulked with mud or bitumen
  - (quffa, 102; hisbiya, 57; ⊕ Tongking boat, 109)
- covered with bark
  - bark canoes, 182
- covered with skin or hide
  - coracle, 111
  - curragh, 133
  - ⊕ Tibetan type, 99
  - elongated
    - umiak, 155
    - kayak, 163

Key and notes for Table 45

if/s	frames inserted; strakes sewn together previously
if/n	frames inserted; strakes nailed, pinned or morticed together previously
pf/n	frames preconstructed; strakes nailed or pinned
sf/n	transom bulkhead 'frames' preconstructed; strakes nailed and clamped

indicates that the craft occurs in the Chinese culture-area

The numbers indicate pages in J.Hornell, *Water Transport; Origins and Early Evolution* Cambridge, 1946.

It should be appreciated that the fundamental characteristic which differentiates Chinese junks and sampans from all other traditional craft throughout the world is the system of transverse water-tight bulkheads, derived as the chart suggests from a natural model ubiquitous in East Asia, the longitudinally split bamboo.

The table is explained in the accompanying pages, and the significance of most of the dotted lines will be clear from the discussion. A few words are necessary, however, on topics peripheral to it.

Thus the balance-board consists of a plank laid athwartships and projecting a considerable distance out to either side. It still exists today off the Somali coast and between India and Ceylon and as far east as Annam. When loaded on the weather side by one or more of the crew, it can give an effective counterpoise to the wind. So far as we know, it was never a Chinese practice. But it was connected with the complex problem of outriggers, i.e. floats of one kind or another, fixed outboard to thwart poles (like balance-boards), and giving stability to craft too narrow or unstable in themselves to attempt long voyages otherwise. The outriggers may be single or double. Their focus of origin, whence they spread by sea as far west as Africa and east to Polynesia, was undoubtedly Indonesia, but the idea itself, according to Hornell, derived probably from some fresh-water technique, and he believed that this was the outboard punting or poling gangway which is an ancient feature of the river ships of South China (Fig. 198). It is simply a light and narrow platform resting on the projecting ends of a number of booms laid at gunwale level athwart the hull. All degrees of intermediate stages are found between this and the double outrigger, among the more interesting being those in which each outrigger carries a number of men paddling. On this view, therefore, the double outrigger derived from the poling gangway of the Indo-Chinese cultural contact zone, as well as from the balance-board of early sea-going Indian Ocean boats; and the single outrigger was a secondary modification which developed at both the oceanic ends of its distribution zone (Madagascar and Polynesia) on account of its greater ability to endure bad weather.

Some craft are known (on Indo-Chinese and South American rivers) which have what might be called 'aerial non-floating out-riggers', i.e. lengths of buoyant material (bamboo, balsa-wood, etc.) lashed alongside the boat but not reaching the water unless the craft heels over when rolling. It is interesting that the *Illustrated Record of an Embassy to Korea in the Hsüan-Ho [Xuan-Ho] reign-period* of 1124 describes cylinders (*tho [tuo]*) of plaited bamboo which were attached to the top of the hull on each side, and which 'gave protection against the waves'. These were on the ships which took the embassy to Korea.

The sailing-raft of logs, or catamaran (from the Tamil *kattu-maram*, tied logs), is connected with the Indian carvel-built ship by a dotted line because there is some reason for thinking that while the oldest carvel-built ships of Europe derived from the ancient Egyptian raft of reeds, some of the boats of India may have derived from wooden rafts. On the coasts of the peninsula (Malabar, Coromandel) there is a tendency to build rafts of odd numbers of logs so fixed that the central one is the lowest, thus approximating to a keel.

reeds or rushes, need not much concern us, for though not unknown in China they were not important for the development of the sailing-craft there. Rafts of logs were widely used by many peoples, especially in the form of the sailing catamaran, so common on the coast of India and the East Indies. In China they are seldom used at sea, but very great rafts of wood still descend the Yangtze [Changjiang] and many of its tributaries. But Chinese wooden rafts were not of importance in the development there of more elaborate shipping; bamboo rafts assuredly were, and to them we shall shortly return.

The moment has now come to describe the basic characteristics of the Chinese junk and sampan, so that its genesis may be more readily appreciated. In Europe and southern Asia the beam forming the keel had a special joint at each end to take another stout beam which turned upwards to form the stempost and sternpost respectively. The long planks forming the hull, which connected them, were held apart to give the desired profile by an internal skeleton of timbers. But junk design, found even in the oldest and least modified types, has a carvel-built hull lacking all three components which were elsewhere regarded as essential – keel, stempost and sternpost. In a junk the bottom may be flat or slightly rounded, and the planking does not close in towards the stern, but ends abruptly, giving a space which would remain open

if it were not filled by a solid crosspiece or transom of straight planks. In the most classical types there is no stem either, but a rectangular transom bow. The hull may be compared to the half of a hollow cylinder bent upwards towards each end, and there terminated by final partitions – like nothing so much as a piece of bamboo slit along its length. Moreover, frames or ribs are replaced by solid bulkheads across the ship (similar to the 'nodal septa' of bamboo itself) of which the stem and stern transoms may be regarded as the outer units.

This is clearly a much firmer method of construction than that found in the ships of other civilisations. Fewer bulkheads were required than frames or ribs to give the same degree of strength and rigidity. It was obviously also possible for these bulkheads to be made watertight, and so to give compartments which would preserve most of the buoyancy of a vessel if a leak should occur, or damage below the waterline. In other ways also the bulkhead structure had various advantages, one example being the provision of the essential vertical support necessary for the appearance later of the hinged rudder. But we shall deal with such matters in due course.

The sampan is reminiscent of the bamboo stem just as much as the junk.

It is an open punt-like boat, wedge-shaped in plan, shallow, keelless, and very broad across the beam; the rail where the deck and sides meet – the gunwale rail – and the strakes are often continued beyond the stern as an upward curving projection, giving the craft the appearance of having cheeks or wings at the stern. It was the roofing of the space between these projections that led to the overhanging stern gallery of the junk.

There have been several theories about the origin of the junk and sampan. One suggests that the design derives from a double canoe in which the two hulls were placed parallel a short distance apart, and connected by planking to form a new bottom with square ends. Craft of this kind, however, have not been found anywhere. Nor is there any evidence of bulkheads along the length of the boat to which these would have given rise. On the other hand, a double canoe as such has existed, and still does, in various parts of the world. And it is curious, too, that the Chinese language contains a number of words to indicate two boats lashed or secured together with cross-timbers side by side. Moreover, such devices are in current use in China, notably for transporting stacks of reeds downstream, and for fishing (the Ichang [Yizhang] 'Water-shoes'). Nevertheless, this line of approach is not convincing.

A modified suggestion is that a process may have taken place like that by which certain fishing boats of Ceylon are still made. A dugout hull is sawn lengthwise into two halves, and these are then connected by frames to which intermediate bottom planks are then nailed. There is, however, no evidence for the use of such a method in China at any time. Indeed if dugout canoes ever did occur in the Chinese culture-area, they were exceedingly rare.

The conclusion to which Hornell came was quite different. He was convinced that we should look to the bamboo raft as the origin of the junk and sampan. In Taiwan the Old World sea-going sailing-raft, he said, attains its highest development, with a 'hull' formed of 9 or 11 curved bamboo poles about 5.5 metres long, strongly bent upwards at the bow end, less so at the stern. Such Taiwanese sailing-rafts are mentioned from time to time in old Chinese literature, especially because of the raids of Taiwanese aborigines on Chinese coastal villages during the twelfth century AD.

If an evolution of the Chinese junk from the bamboo raft form is envisaged, it is only necessary to suppose a conversion of the wooden cross-beams into bulkheads, the substitution of wood planks for bamboo in bottom and sides, and the addition of decking. Such a process can actually be seen at work in the catamaran log rafts of Madras, some of which have plank strakes pegged on along each side. In many Chinese vessels, notably the Liu-phêng chhuan [Liu-peng chuan] of Kuangtung [Guangdong], the lines of the sailing raft are preserved and exaggerated.

It is not necessary to insist upon the sea-going bamboo sailing raft of Taiwan as the only ancestor of all junks, for many other forms of bamboo raft

exist, and are still regularly in use on Chinese rivers. One of the most interesting is the Ya River raft of Szechuan [Sichuan], which moves up and down 160
km of intractable waterway between Yachow [Yazhou] and Chiating [Jiading],
carrying Tibetan trade. It must be one of the lightest-draught general cargo-
carriers in the world, for its depth below the water-line when loaded (with a
cargo of 7 tonnes) is often no more than 8 centimetres and never exceeds 16,
owing to the buoyancy of the bamboo. In length the rafts, which are quite
unsinkable, vary between 6 metres and 33 metres and are built throughout of
the stems of the giant bamboo (*Dendrocalamus giganteus*) which grows as high
as 24 metres with a diameter of as much as 30 centimetres. The bow is
narrowed and bent upwards in a curve by heating, so that the raft can slide over
rocks which may be almost level with the water (Fig. 178). In other provinces
there are also interesting bamboo rafts, and some of these also have the
upturned bow. Moreover, certain boats, such as the 'fan-tail' of western
Szechuan [Sichuan], seem to have transferred this ancient device to the stern
as a protection against shipping water when descending rapids.

What Hornell never knew was that there are indigenous Chinese traditions that the junk was developed from the raft. The *Shih I Chi* [*Shi Yi Ji*]
(Memoirs on Neglected Matters) of the third or fourth century AD says:
'Hsien-Yuan [Xian-Yuan] changed the custom of floating on rafts (*fu*), for he
invented boats and oars.' To which a dictionary compiler remarked 'So that
before there were boats, people crossed rivers by means of rafts. Since the
term *fu* means the same thing as *fa*, these rafts must have been known before
(the legendary emperor) Huang Ti's [Huang Di's] time'. Moreover, the view
that the junk developed from the bamboo raft is not contradicted by the
famous passage in the *I Ching* [*Yi Jing*] (Treatise of Art and Games) of the
third century AD, where it is said of the sages that they 'hollowed out logs to
make boats, and hardened wood in the fire for oars'. This usual translation lays
too much weight on the meaning of the first word, which can also signify to rip,
to cut off a slice, to cut up into several parts, to cut up an animal, and so on.
Here it could equally refer to cutting a log into planks. And as we shall see
before long, the ancient pictogram for a boat shows ends which are square and
not pointed. Moreover, it may be significant that several ancient books refer to
the use of large bamboos for making boats.

CONSTRUCTIONAL FEATURES OF THE JUNK AND SAMPAN

The best way to proceed from the point we have now reached will be to
examine more closely a few typical examples of marine architecture. At the
same time we shall be able to gain an idea of some of the most important
technical terms which were, and are, used by shipwrights and sailors.

Fig. 178. A bamboo raft from the *Thu Shu Chi Chhêng* [*Tu Shu Ji Cheng*] (Imperial Encyclopaedia) of AD 1726, redrawn from *San Tshai Thu Hui* [*San Cai Tu Hui*] (Universal Encyclopaedia) of 1609. Note the bamboo side-rails, which could be trusses to prevent sagging of bows and stern.

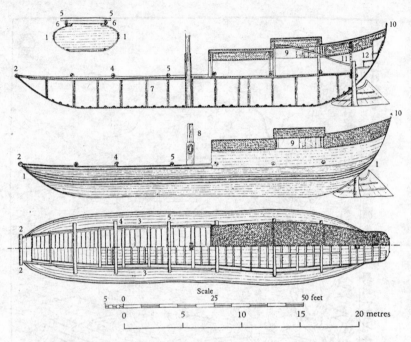

Fig. 179. A river-junk of the Upper Yangtze [Yangzi], the Ma-yang-tzu [Ma-yang-zi], here taken as a prototype of all Chinese shipping (after G.R.G. Worcester).

1   long and heavy wales running the length of the vessel

2   projecting cross-beam at transom bow

3   low raised borders to prevent water on deck from running into hatches, and stretching from bow to deckhouse

4, 5   projecting cross-beams holding the thole-pins (pins against which the oars press as a fulcrum) for the self-feathering oars or yulohs.

6   transoms

7   fifth bulkhead supporting hardwood capstan

8   24 metre pinewood mast and its tabernacle (elevated socket) with carved cleat for the ropes

9   tiller-room with forward view (but in some forms the tiller may rise above this deckhouse and be worked from a crosswise gangway above it)

10   two tall pins on which bamboo rope is coiled

11   7.6 metre tiller of balanced rudder

12   after-house (skipper's cabin and home)

Type-specimens

Let us begin with a cargo-boat of the Upper Yangtze [Yangzi], the Ma-yang-tzu [Ma-yang-zi], Figs. 179 and 180. Like all river junks it is quite variable in size, from bow to stern measuring between some 11 and 30 metres. Formerly they were built as large as 46 metres. As will be seen from Fig. 179,

Fig. 180. Model of a river-junk, a Ma-yang-tzu [Ma-yang-zi] in the Science Museum, London.

there are no less than 14 bulkheads, forming so many separate holds. In the oldest and most characteristic build there is no basic lengthways straightening timber (i.e. keel) at all, the structure depending for rigidity only on the planking nailed to the bulkheads, and on very solid 'wales' (thick strakes) along the sides. These are still fitted, taking their place among the ordinary strakes, but modernisation has now sometimes induced a 'keel' even in remoter river types of build. Or inside the planking there may be a 'kelson' (a line of timber

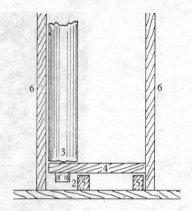

Fig. 181. Tabernacle or elevated socket of a typical Chinese mast (after G.R.G. Worcester). A tenon (2) On the heel (3) of the mast fits into a socket in a horizontal timber (4) bearing on the ribs (5) and fitting snugly against the sides of the bulkheads (6). This prevents the heel of the mast moving forward, and helps to distribute the thrust.

placed inside the ship along and above the floor timbers and parallel to the keel) or two side kelsons at the turn of the 'bilges' (the places where the ship's hull turns towards the keel).

Between the bulkheads, which frequently are stiffened, there may be some frames, half-frames, or ribs, though it is doubtful if these are of ancient origin. Floorboards may lie above the structural bottom planking. Many types of Chinese craft have considerable 'tumblehome' (inward inclination of the upper sides), like European ships of the eighteenth century; in other words they are 'turret-built'. The deck does not, therefore, occupy by any means the whole breadth of the ship. As is almost invariably the case with Chinese ships, all superstructures lie aft of the main mast, thus giving uninterrupted space for handling oars, tracking gear and so on. The deck is carried on beams across the breadth of the ship and placed at intervals along the top of the hull. Some of these project beyond the sides and serve as pivot points for different kinds of oars, so that these may clear the sloping upper boarding of the hull. The square bow of the ship ends in a massive projecting cross-beam useful for all kinds of purposes. Raised boarding round the hatches prevents water on deck from pouring down on to the cargo in the holds.

The pinewood mast, some 24 metres in height, is fitted in a tall elevated socket or 'tabernacle' (Fig. 181), rising perhaps a couple of metres above the deck. The mast carries a lofty lug-sail (see p. 185). From Fig. 179 it will be seen that the rudder is a balanced one (that is to say, part of its blade lies in front of the axis about which it pivots). It carries a tiller some 7.5 metres long, which may require as many as three men to handle in a difficult rapid. While the permanent crew of a large river junk of this kind may only be 8, 50 or 60 more men will be engaged from time to time; and for towing upstream in certain places, as many as 400 may be necessary.

As our type-specimen of the sea-going junk we may take the Chiangsu [Jiangsu] freighter or 'sand ship'. In the past these reached a length of as much as 52 metres. The pinewood hull is flat-bottomed (see Fig. 182), the central lengthwise timber being somewhat larger than the others and substituting for a keel. As many bulkheads are present as in the up-river boat just described, and the sides of the hull are strengthened by wales. Since the curves of the hull meet at bow and stern, the foremost and after compartments are masterpieces of construction, and the curved deck beams are tongued and grooved with great ingenuity into the curved frames of the hull. What is more, certain of the prow and stern timbers are actually grown to shape. Both bow and stern are 'bluff' (flattened) and capable of withstanding the worst weather, while as the drawing shows, there is a kind of 'false stern' at the very rear. This is constructed by extending the sides of the hull in a rising curve beyond the final transom (cross-piece) and ends in a shorter false transom about 2 metres above the water-line. The decked surface of this structure prolongs the deckhouses and carries a windlass for hoisting or lowering the rudder, which is enclosed within this space. The rudder-post itself works in three open-jawed wooden pivots, and the tiller is handled either on the roof of the deckhouse or from inside it. Such an arrangement has been for centuries particularly characteristic of Chinese ships. As will also been seen from Fig. 182, further aft of the false stern is a long stern-gallery.

The ship has five masts, a fact which (as we shall see) seemed surprising to the Europeans of the thirteenth century, and appears to have exerted a great effect on their subsequent ship design. Generally speaking, all sea-going junks of any size were, and are, provided with multiple masts, river craft rarely having more than two. A feature, however, which did not spread outside the Chinese culture-area, was the system of staggering the masts to port (left) and starboard (right) positions. Thus in the present case, the foremast is placed off-centre to port, the second foremast is amidships, and both are tilted or 'raked' forward. The main mast is also amidships, but raked slightly backwards. Next comes the mizen mast; this is on the port side with a marked forward rake. Lastly there is the 'bonaventure' or aft mizen mast, considerably taller than the mizen itself and with no rake at all. The raking is not the same on all ships, but the general tendency is to have the masts radiating like the sticks of a fan.

There is individual variation in the construction of the tabernacles to hold the masts and, as in nearly all traditional Chinese sea-going ships, the masts are completely devoid of stays. However, some of the heavy primary masts are provided with single or Y-shaped struts at about deck level; these

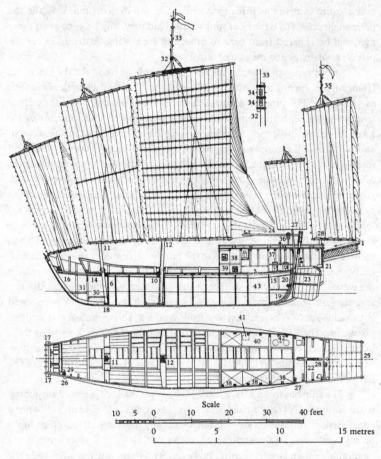

Fig. 182. Chiangsu [Jiangsu] freighter, a sea-going junk which was probably
the parent type for many kinds of Chinese ships (after G.R.G. Worcester). It
was often nearly 60 metres in length, approaching the size of the great wooden
ships of the Ming navy, but the example here drawn is of 26 metres.

1	block and multiple sheets (ropes) attached to the aft mizen-sail
6	first internal bulkhead
10	one of numerous strong deck beams
11	14 metre midship foremast (forward rake or tilt)
12	21 metre mainmast (slight aft rake)
14	fore compartment, usable for living quarters and storage of ropes for the rigging
15	after compartment, abaft of the twelfth and last internal bulkhead
16	three fore-and-aft stem ribs grown to shape
17	bow cross-beam
18	heel of foremast
19	for-and-aft stern timbers grown to shape
20	blunt and rounded transom stern

transfer some of the thrust from the sail to the junctions of hull and bulkheads (Fig. 183). Such junks as these, which today carry a crew of about 20, must resemble fairly closely the prototypes which journeyed to the Indian Ocean in the Sung [Song] period. Their beauty when in full sail has fascinated many observers.

Technical terms

As already mentioned, the terms used in Chinese shipbuilding and nautical matters are not easy to elucidate. So far as is known there is nothing in Chinese literature that sets out such terms; practical men did not commit themselves to writing, and literary men had little or no knowledge of the building and handling of ships. At best scholars could only make commentaries on technical terms which even their predecessors had perhaps only half understood. So although Chinese encyclopaedias from the third century BC onwards, generally contain sections devoted to shipping terms, it is noticeable that the majority of these concern types of boats and ships long obsolete or not easily identifiable. The number of terms for distinct parts of a ship and its gear is smaller. Even then, much space is devoted to the identification of dialect phrases or local usages, so that the task of selecting out items of information which really prove the existence of any given technique at any particular time in history is going to require prolonged research.

Western sinology is not much more helpful. Even the copious and careful works of the contemporary scholar W.R.G. Worcester do not give us

Caption for fig. 182 *(cont.)*

21	false stern, consisting of a 2.4 metre extension of the sides of the hull beyond the transom, ending in a shorter false transom
22	windlass for hoisting the rudder
23	rudder, iron-bound and non-balanced
24	4.9 metre tiller
25	3 metre stern gallery
26	9.4 metre port foremast (forward rake)
27	8.5 metre port mizen mast (forward rake)
28	14.6 metre 'bonaventure' or aft mizen mast (vertical), slightly port of amidships
29	port foremast tabernacle (inside bulwarks)
30, 31	fore-and-aft baulks securing midship foremast heel
32	hounds or cheeks of mainmast
33	light topmast of mainmast
34	sheave (pulley) pins passing through both masts and securing double halyard sheaves
35	light topmast of aft mizen mast
36	navigation light
37	galley
38, 39	cabins with bunks and sliding doors
40	rice bin and stores
41	shrine to Kuan Yin (Goddess of Mercy)
42	cooking-stoves
43	below-deck living quarters

Fig. 183. Deck of Swatow freighter. The mainmast strut, which transfers part of the thrust of the wind on the sail to the hull and bulkheads forward, is prominent. One of the usual iron bands and wedges on the mast can also be seen, and the halyard winches (one dismounted) on each side to port and starboard. From the Waters Collection, National Maritime Museum, Greenwich.

the Chinese equivalents for every technical term used. This is because he met with unexpected difficulties since, in general, even the finest shipwrights and sea-captains with whom he worked, could not write, nor could any members of their crews or families. It is clear, then, that there are certainly many spoken craft terms for which no written forms exist at all. Indeed, an official eighteenth-century Chinese scribe had to invent characters for certain words used by his informants. Moreover, the Chinese seamen seem never to have troubled to elaborate that infinity of technical terms covering the smallest part of the gear, in which Europeans have delighted. And lastly terms vary from port to port.

Though it is not specifically a treatise on shipbuilding, a great mass of information (not yet digested by historians) is nevertheless to be found in the book on the Yangtze [Yangzi] shipyards, the *Lung Chiang Chhuan Chhang Chih* [*Long Jiang Chuan Chang Zhi*] (Records of the Shipbuilding Yards (near Nanking [Nanjing]) on the Dragon River) of AD 1553 by the nautical writer Li Chao-Hsiang [Li Zhao-Xiang]. We shall return to this at a more convenient place (p. 124). The best picture so far found in a Chinese work is contained in the *Account of the Liu-Chhiu* [*Liu-Qui*] *Islands* by Chou Huang (Zhou Huang) already mentioned (p. 61), and is shown in Fig. 184. It is particularly valuable because the artist has added a number of technical terms as can be seen in the caption accompanying our reproduction. Its four masts are stepped with the characteristic Chinese sails which are well drawn, while it is clear that ad-ditional sails, and masts or spars for carrying them, as in Marco Polo's time, were present. The combination of sails illustrated here is not dissimilar in principle to that used in modern racing practice and contrasts with the more limited usage of the 'full-rigged ship' of Renaissance Europe. Our junk is obviously running merrily before the wind. Other points to be noted are the use of deck winches for hoisting sail, and the slung rudder, partially raised to reduce water-resistance, as well as the 'bousing-to tackle' which runs from the foot of the rudder, under the bottom of the vessel, and thence to a windlass in the forecastle thus holding the rudder against its wooden jaws. It will also be seen that the ship has two anchors at the bow and one without a cross-bar (or stock) on the port side, and that it also possesses portholes.

While there exists no great published work on Chinese shipbuilding, some manuscript material is available, and there is probably a lot more still lying in Chinese provincial archives. Of manuscripts in Europe, that at the Marburg Library has already been mentioned (p. 61). Entitled *Min Shêng Shui-Shih, Ko Piao Chen Hsieh Ying, Chan Shao Chhuan Chih Thu Shuo* [*Min Sheng Shui-Shi, Ge Biao Zhen Xie Ying, Zhan Shao Chuan Zhi Tu Shuo*] (Illustrated Explanation of the (Construction of the Vessels of the) Coastal Defence Fleet (Units) of the Province of Fukien stationed at each of the Headquarters of the several Grades), it is important because it contains some sixty drawings of the component timbers of five classes of ships. Since the component parts are labelled rather clearly as Fig. 185 shows, it is possible to confirm a number of important words. However, Chinese literature does contain several interesting accounts of the construction of model ships for instructional purposes. About AD 1158, Chang Chung-Yen [Zhang Zhong-Yan] was in the service of the Jurchen Tartar dynasty. Of him we read:

When they began to build ships, the artisans did not know how. So Chung-Yen [Zhong-Yan] made with his own hands a small boat several (tens of) centimetres long. Without the use of glue or lacquer

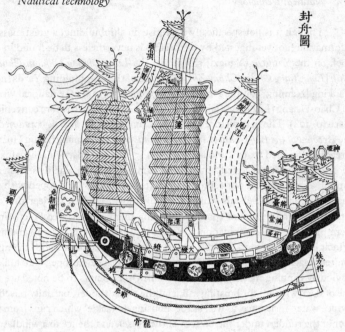

封舟圖

Fig. 184. The ocean-going junk in the *Liu-Chhiu Kuo Chih Lüeh* [*Liu-Qiu Guo Zhi Lüe*] (Account of the Liu-Chhiu (Liu-Qiu) Islands) of AD 1757, one of the best pictures of a Chinese ship in the Chinese style to be found in literature.

fêng chou [*feng zhou*]   government ship

thou po [*tou bo*]   'kerchief' or 'headcloth', spinnaker (sail)

thou chhi [*tou qi*]   'pursuer', water-sail or bowsprit-sail. In the West the bowsprit sail is a descendant of the Roman *artemon* sail. The *Santa Maria* of Columbus (1492) had it, but it did not become general on Western ships until the end of the sixteenth century.

mien chhao phai [*mian chao pai*]   'avoidance of courtesy' notice, i.e. 'bound on important government business'

thou phêng [*tou peng*]   foremast mat-and-batten sail

phêng khu [*peng ku*]   'trousers', foot of foresail, or perhaps foresail boom

ting [*ding*]   grapnel anchor

tu lê [*du le*]   bousing-to tackle securing the rudder

lung ku [*long gu*]   'dragon spine', central longitudinal strengthening member of hull

erh liao [*er liao*] ⎫ winches ('winders') for hoisting sail
ta liao [*da liao*] ⎭

phêng chhün [*peng qun*]   'skirt', foot of mainsail, or perhaps mainsail boom

ta phêng [*da peng*]   mainsail (of bamboo matting and battens)

chha hua [*cha hua*]   inserted ensign

thou chin ting [*tou jin ding*]   topsail (of cloth or canvas)

i thiao lung [*yi tiao long*]   dragon ensign

shen chhi [*shen qi*]   'spirit flag'

wei sung [*wei song*]   mizen-sail

chen fang [*zhen fang*]   compass cabin

shen thang [*shen tang*]   chapel

Chiang thai [*jiang tai*]   poop

shen têng [*shen deng*]   'spirit light'

thieh li tho [*tie li tuo*]   ironwood rudder

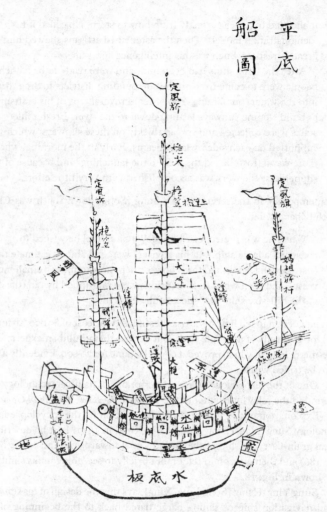

Fig. 185. Drawing from the most important manuscript shipbuilding manual
Min Shêng Shui-Shih . . . Chan Shao Chuan Thu Shuo [*Min Sheng Shui Shi . . .
Zhan Shao Zhuan Tu Shuo*] (Illustrated Explanation of the (Construction of
the Vessels of the) Coastal Defence Fleet (Units) of the Province of Fukien
[Fujien] stationed at each of the Headquarters of the several Grades),
preserved in the Prussian State Library at Marburg. Though the manuscript
may be of the middle nineteenth century, the contents belong to the first half
of the eighteenth. This 'flat-bottomed ship' bears considerable resemblance to
that in Fig. 184. Notice the 'weather-vane pennants (*ting fêng chhi* [*ding feng
qi*]), the 'standard of (the goddess) Ma-Tsu [Ma-Zu]' at the stern, the rudder
(*tho* [*tuo*]) and sculling-oar (*lu*), the oculus or eye (*lung mu* [*long mu*]), and the
bulwark gate (*shui hsien mên* [*shui xian men*]).

it all fitted together perfectly from bow to stern. He called it his 'demonstration model'. Then the astonished artisans showed him the greatest respect. Such was his intelligence and skill.

After the large ships had been built and were ready to be launched, people were to come from all the surrounding districts to drag them into the water, but Chung-Yen ordered several tens of his craftsmen to build sloping runways leading down to the river. Fresh millet stalks were collected and spread thickly on these slipways, which were supported on each side by large beams. Early in the morning, when there was a frost, he led the men to the launching, and because of the slipperiness the work was accomplished with very little effort.

A contemporary in the service of the Sung [Song] down south was Chang Hsüeh [Zhang Xue]:

> When he was prefect of Chhuchow [Chuzhou] he wished to construct a large ship, but his advisers were not able to estimate the cost. Hsüeh [Xue] therefore showed them how to make a small model vessel, and then when its dimensions were multiplied by ten (the cost of the full-size ship) was successfully estimated.

The text goes on to tell how his artisans estimated the cost of 80 000 'strings of cash' for the walls of a temple park, but he had them build an experimental three-metre length, which proved it could be done for 20 000. Evidently a man not to be trifled with.

One of the few serious literary descriptions of nautical technology is to be found in the relevant chapter of the *Thien Kung Khai Wu* [*Tian Gong Kai Wu*] (Exploitation of the Works of Nature) written by the encyclopaedist of technology Sung Ying-Hsing [Song Ying-Xing] in 1637. In it he describes a typical grain-carrying ship of the Grand Canal towards the end of the Ming (Fig. 186) and then goes on to refer more briefly to sea-going junks similar to that shown in Fig. 182.

Sung Ying-Hsing [Song Ying-Xing] says that the design of his standard inland navigation ship or sailing barge dated back to the beginning of the fifteenth century AD, in the Yung-Lo reign-period. On account of grain-transport losses on the sea route, it was decided to revert once more to the use of the Grand Canal. Accordingly, the present shallow-draught canal boats were introduced by a certain Mr Chenn [Zhenn], lord of Phing-Chiang [Ping-jiang]. Sung [Song] then continues:

> ... The general construction of a canal ship is as follows: a bottom (of stout planking) serves as the foundation, there are (across the ship and fore-and-aft) timbers like the walls of a building, and there is bamboo tiling (to cover the hold) as if it were a roof. (The compartments)

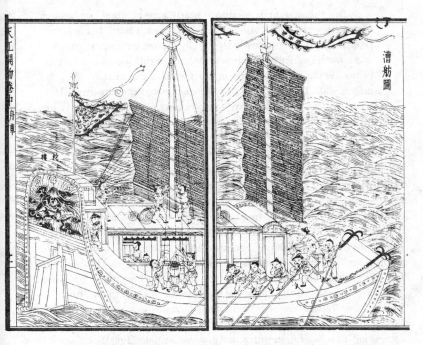

Fig, 186. A grain freighter of the Grand Canal at the beginning of the Ming; from *The Exploitation of the Works of Nature*, 1637.

forward of the mast framework (j.e. the tabernacle and its associated structures) are like the main gates, and (the compartments) aft of it are like the sleeping quarters. The mast is like (the stock of) a crossbow, and the halyards [ropes for raising or lowering the sails or 'yards'] and sails are like wings. Oars (may also be motive power) as the horse is to the cart; hauling cables are as the shoe is to the walker. The cordage adds strength to the bones like the bones and sinews of a hawk. The bow-sweep [a large oar] goes before like a spearhead, the rudder (at the stern guides the direction of the vessel) like a commander, and the anchors call a halt like an army encamping for the night.

On the original specification, the grain-ship is built 16 metres long with planks 5 centimetres thick. The choicest timber for it is large baulks of *nan-mu*, but chestnut is also used as second best....

The general plan is thus uncompromisingly flat-bottomed and keel-less. It also has numerous transverse bulkheads, each of which Sung [Song] specifi-cally names.

Such a ship could take a cargo of nearly 2000 piculs of rice [though
in practice only 500 piculs were delivered to each boat], but another
type designed independently later on by the army transport service
was 6 metres wider both at bow and stern, so that it could carry 3000
piculs. As the flash-lock gates in the (Grand) Canal are only 3.6
metres wide, these craft are just slim enough to negotiate them. The
boats used nowadays for travelling officials are of just the same type
(as the freighters), but their windows, doors and passageways are
made somewhat wider, and besides they are more elegantly painted
and finished; that is all.

It is interesting to compare these figures with others for the nineteenth
century. A burden of 2000 piculs is about 140 tonnes. According to the scholar
G.M.H. Playfair, 670 grain junks in 1874 transported to the capital 1.36
million piculs of grain (96 000 tonnes), so that the lading was about 143 tonnes.
The average size of the ships thus remained at that time just about what it had
been at the beginning of the seventeenth century, and indeed the beginning of
the fifteenth. From the evidence of the size of Chinese river-locks this tonnage
had clearly been attained very nearly by the eleventh century. But it had
probably dropped greatly during the three centuries when the sea route was
predominant. Indeed, just after mentioning the 2000 piculs, Sung Ying-Hsing
[Song Ying-Xing] adds, in one of his 'footnote' commentaries, that upon
comparing his specification with actual practice, he found that the burden was
generally not more than 500 *tan* [*dan*] (about 35 tonnes). He next goes on to tell
what he saw on his visits to shipyards:

> The construction of a boat begins with the bottom. The strakes of
> the hull are built up on both sides from the bottom (planking) to a
> height (equivalent to that of the future) deck. Bulkheads are set at
> intervals to divide the vessel (into separate compartments), and (the
> holds have) sheer vertical sides. . . . The hull is covered at the top (or,
> surmounted) by great longitudinal members. The (winches for the)
> halyards are fixed above these. . . .   . . .

We have a number of eye-witness accounts of Chinese shipbuilding in the
present century, all confirming and elaborating what Sung had to say. Many
observers have been struck by the fact that the Chinese traditional shipwrights
used no templates or blueprints, depending rather upon the skill and sureness
of eye of the oldest and most experienced craftsmen. Though some technical
manuals have existed, the greater part of the industry must always have been
based on the personally transmitted 'know-how' of the masters. Elsewhere in
the shipyards, Sung sees the finishers at work:

> The joints between the planks are caulked by first forcing in
> tangled jute filaments with a blunt chisel. Then a (putty-like)

composition of fine sifted lime and tung oil (is applied to complete the job). In Wênchow and Taichow, and in Fukien and Kuangtung [Wenzhou, Daizhou, Fujian and Guangdong], the ash of oyster-shells is used in place of lime.... For sea-going ships the caulking is done with a mixture of fish oil and tung oil, why, I do not know.

Then he has a look at the stores:

The timber for the mast is usually fir, which must be straight and sound. If the natural size of the spar is not long enough for the mast, two pieces can be coupled together by means of a series of iron bands placed around the joint a few centimetres apart. An open space is left in the deck for the mast. For the stepping of the main-mast, its upper part is laid across several large boats brought alongside, and the top hoisted up (into position) with a long rope.

Hull timbers and bulkheads are made of *nan-mu*, *chu-mu* [*zhu-mu*], camphor wood, elm and sophora wood. [Camphor wood, if taken from a tree felled in spring or summer, is liable to be attacked by boring insects or worms. (Sung's own commentary.)] Deck planks can be made of any wood. The rudder post is made of elm, or else of *lang-mu*, or of *chu-mu* [*zhu-mu*]. The tiller should be of *chu-mu* or *lang-mu*. The oars should be of fir or juniper, or catalpa wood. These are the main points.

Sheets (sails) and halyards carried are made of prepared hemp fibres twisted roughly together until they reach a diameter of some 3 centimetres; then the rope can sustain a weight of 10 000 *chün* [*zhun*] (about 180 tonnes). Anchor cable is made of thin strips of the outer parts of the stems of green bamboo, which after being boiled in water are twisted into rope. The tracking (towing) cables are made in the same way. Cables more than 30 metres long come in sections with loops (or eyes) at both ends to join them together. It is in the nature of bamboo to be 'straight' (i.e. to have a high tensile strength), so that one such skin strip can sustain a weight of 1000 *chün* (18 tonnes). When ships are going up the Yangtze [Yangzi] gorges to Szechuan [Si-Chuan] they do not use the twisted cables, but rather bamboo laths cut to the width of some 2 centimetres and joined to form a long flexible spar (or chain of rods); this is necessary as twisted cable can easily be cut or broken by the sharp rocks.

Finally he makes some remarks about the general rig of the vessels of the inland waterways:

Ships more than 30 metres long must have two masts. The main-mast is stepped two bulkheads forward of the mid-point, and the fore-mast some 3 metres or more further forward. On the grain ships the main-mast is about 24 metres high, but spars shorter by one or two tenths of this are also used. The heel (bottom of the mast) is steeped into the hull by a depth of some 3 metres. The position (of the halyard blocks) of the sail is at a height of 15 to 18 metres. The fore-mast is less than half the height of the main-mast, and the dimensions of the sail which it carries are not much more than about one third (of the main sail). Rice transport boats in the six prefectures of Huchow [Huzhou] and Suchow [Suzhou] have to pass under stone bridge arches, along waterways without the dangers of the great rivers like the Yangtze and the Han, so that the dimensions of their masts and sails are much reduced. Ships which navigate in the provinces of Hunan, Hupei [Hubei] and Chiangsi [Jiangxi], crossing the great lakes and rivers, amidst incalculable winds and waves, have to have their anchors, cables, sails and masts well found according to the standard regulations exactly – then there need be no anxiety.

About sea-going ships Sung Ying-Hsing [Song Ying-Xing] has little to say (they were evidently peripheral to his experience), but he recorded some interesting names. Among the sea-going junks used for grain transport during the Yuan dynasty and the beginning of the Ming, there were the 'shallow-draught ocean ship', the 'boring-into-the-wind ship' [or the 'sea-eel' or sea-serpent' ship]. Of these, he says:

... Their voyages did not exceed 10 000 *li* (some 5350 km) along the coasts, across the Dark Sea and past the Sha-mên [Sha-men] islands. No great danger was encountered in the voyage. As compared with the junks that sailed to Japan, the Liu-Chhiu [Liu-Qiu] islands, Java and Borneo for trade, these (coastal transports) were not one-tenth as large and expensive.

The build of the first type of sea-going junk is similar to that of the canal junk, save that it is some 5 metres longer, and 75 centimetres broader in the beam. All else is the same except that the rudder-post must be made of iron-wood (a very hard wood)....

(All) the ships that sail to foreign countries are rather similar in specification to (the great) sea-going junks (just mentioned). Those which hail from Fukien [Fu-Jian] and Kuangtung [Guangdong] ... have bulwarks of half-bamboos for protection against the waves. The ships from Têngchow [Dengzho] and Laichow [Laizho] (in Shantung [Shandong]) are of different type again.... All these types have in common two mariner's compasses, one at the bow and the other at the stern, to indicate the direction of the course. They also have in common 'waist-rudders' (i.e. 'leeboards' or strong plank frames let down into the water from the side of a ship to prevent a sideways drift to leeboard – the side sheltered from the wind) . . . They carry several piculs of fresh water, enough for the whole ship's company for two days, in bamboo barrels, and when they touch at islands they get more.

Everything that has been said in the preceding pages about the construction of the junk is applicable, having regard to size, to the larger and smaller sampans whether decked or otherwise. These have won the admiration of the sailors of other cultures for their excellent adaption to their fishing, or freight-carrying duties, no less than the largest and most majestic traditional Chinese vessels.

Hull shape and its significance

The ship which, when cut across the middle, had a shape like a rectangle with rounded corners, certainly had the future before it. This is the form of iron and steel steamships of our own time, and it was already apparent in Europe more than a century and a half ago that such a shape possessed very great advantages, among them stability when loading and unloading. But Chinese junks, which were also of this shape, had another characteristic that surprised Europeans at the first encounter, yet which had also been adopted by them (if in less extreme forms); this was a build in which the broadest part at water level lies aft of the midship line.

The ways of moulding the hull of a ship are limitless, but a simple distinction can be made between three basic forms. The first are those which are symmetrical, their greatest area lying amidships; the other two are either broader forward or broader aft. Such broadness may be obtained by varying the way the hull tapers, increasing it gradually in one direction more than in

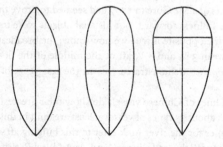

the other. Broadly speaking, the European tendency has always been to set the greater fullness of the ship forward, towards the bow, while the Chinese tendency was to set it towards the stern. From parallels drawn by European writers in the seventeenth century, it appears clear that in the West it was felt that a ship ought to imitate the shape of a fish, whereas the Chinese were of the opinion that a ship should be shaped like a the outline of a swimming bird at water level. Admiral Paris in 1840 was perhaps the first observer to note this great difference. In his *Essai sur la Construction Navale des Peuples Extra-Européens*, he wrote:

> ... For our best hulls we have taken the fishes as models, always larger at the cephalic end, but the Chinese, who also copied Nature, imitated the palmipeds (the web-footed birds), which float with the greatest breadth behind, for somewhat obscure reasons. In this they were acute, for aquatic birds, like boats, float between the two media of air and water, while fish swim only in the latter. These strange people seem to have done everything in the opposite way to which it is done at the other end of the continent, and they copied Nature still further in seeking to exert the greatest possible propulsion on the stern, instead of applying, as it were, a tractive force to the prow. This led them to the use of those strong paddles (the 'yulohs') which imitate in position the web-feet of the palmipeds – a position which must have been very important for swimming, since it deprived such birds of the facility of walking easily on land, and even in the case of the best swimmers, prevented this altogether. These very simple observations (which the Chinese have utilised) will find one day, perhaps, a happy application to the steam-boat, which, set in motion as it is by an internal force not coming like the wind from without, finds itself in exactly the same situation as the swimming bird, and might gain from a closer approximation to the latter's form.

These prophetic words of Paris were justified within the next twenty years by the coming of the screw propeller. As for the hull shape, the pioneer water-tank experiments of a Mr Gore in 1797 had seemed to prove the superiority of shapes with the fullness forward, but his trial objects were totally immersed. That precisely the opposite is true we now know; the broadest section should be located between 3 % and 8 % aft of the middle of the flotation line. The racing yacht *America* demonstrated this in the last years of the nineteenth century.

That the lines of Chinese vessels do show their greatest fullness aft has been confirmed; about 35 types of vessel at least are built in this way. Several of the familiar names for the river junk refer to this bulging aft – the 'hempseed oil bottle', the 'ingot', the 'red slipper', and, as Admiral Paris would have been delighted to learn, the 'duck's tail'.

Chinese shipwrights were in fact quite well aware of what they were doing. In his *Hai Yün Hsin Khao* [*Hai Yun Xin Kao*] (New Investigation of

Sea Transport) of 1579, Liang Mêng-Lung [Liang Meng-Long] alludes to a famous shipbuilder of the Yuan period, Lo Pi [Lo Bi], who constructed junks with the significant name of Hai-phêng [Hai-peng]. The reference was to the gigantic bird which could take off from the water and fly to the south. Lo Pi's [Lo Bi's] motto was 'with the body like a tortoise and the head like a snake'. No more definite statement could be imagined.

Water-tight compartments

There is another valuable shipbuilding technique which Europeans adopted from China, and that is the water-tight compartment. From all that has gone before it will have become evident that bulkhead construction was the most fundamental principle of Chinese naval architecture (Fig. 187), and water-tight compartments were its consequence. This is one of those cases where it is also possible to be fairly precise as to the date when the method was adopted in the West. It was 'in the air' during the closing decades of the eighteenth century, and nearly every reference of the period recalls its use in Chinese ships. Indeed, in 1795 Sir Samuel Bentham had been commissioned by the Admiralty to design and construct six vessels of entirely new type, and he built them with 'partitions contributing to strength, and securing the ship against foundering, as practised by the Chinese of the present day'.

The circumstances of Bentham's introduction are so interesting with regard to the diffusion of techniques that they are worth a closer look. From Lady Bentham's biography of him, which drew upon his early journals, we find that though in the latter part of his life he was for long chief engineer and architect of the British Navy, his earlier years had been spent in the Russian service. In the course of extensive Siberian travels in about 1782, he reached the frontiers of China at Kiachta where he met the local Chinese governor and observed large boats on a river which must have been the Shilka. Some of these, if not all, would have had the usual bulkheads. Lady Bentham wrote:

> This was no invention of General Bentham's; he has said himself officially that it was 'practised by the Chinese of the present day, as well as by the ancients', yet to him is due the merit of having appreciated the advantage of water-tight compartments, and of having introduced the use of them.'

Still more interesting is the fact that Bentham copied another Chinese technique. About 1789, when serving as colonel of a regiment in the Crimea, he constructed an articulated boat, the *Vermicular*, for negotiating the shallow and winding reaches of certain rivers down which natural products were to be brought. As we shall shortly see (p. 96) this was a characteristic Chinese practice.

The only remarkable thing about this transmission of the water-tight

Fig. 187. Photograph of the port quarter of a South Chinese freighter under repair in the shipyards of Hongkong. Four or five bulkheads are seen because of the removal of the strakes of the hull. Within the upcurving stern itself, an additional five ribs or frames are visible. The slot for the slung rudder can just be made out under the overhang of the stern gallery. From the Waters Collection, National Maritime Museum, Greenwich.

compartment is that it should have taken so long to be adopted in the West. Marco Polo distinctly described such compartments in Chinese ships about 1295, and his information was repeated by the merchant Nicolò de Conti in 1444.

Less well known is the intriguing fact that in some types of Chinese craft the foremost (and less frequently the aftermost) compartment is made free-flooding. Holes are purposely drilled in the planking. This is the case with the salt-boats which shoot the rapids down from Tzuliuching [Ziliujing] in Szechuan [Sichuan], the gondola-shaped boats of the Poyang [Boyang] Lake, and many sea-going junks. The Szechuanese boatmen say that this reduces resistance to the water to a minimum, and the device must surely cushion the shocks of pounding when the boat pitches heavily in the rapids, for she acquires and discharges water ballast rapidly just at the time when it is most desirable to counteract buffeting at stem and stern. The sailors say that it stops junks flying up into the wind. It may be the reality at the bottom of the following story, related by Liu Ching-Shu [Liu Jing-Shu] of the fifth century AD in his book *I Yuan* [*Yi Yuan*] (Garden of Strange Things):

> In Fu-Nan (Cambodia) gold is always used in transactions. Once there were (some people who) having hired a boat ... had not reached their destination when the time came for the payment of the pound of gold which had been agreed upon. They therefore wished to reduce the quantity (to be paid). The master of the ship then played a trick on them. He made (as it were) a way for the water to enter the bottom of the boat, which seemed to be about to sink, and remained stationary, moving neither forward or backward. All the passengers were very frightened and came to make offerings. The boat (afterwards) returned to its original state.

This, however, would seem to have involved openings which could be controlled, and water pumped out afterwards.

One reason for using water-tight compartments with a free-flooding portion is to enable fishing-boats to bring their catch to port and market in live condition. This was easily done in China, but the practice was also known in England, where the compartment was called a 'wet-well', and the boat in which it was built, a 'well-smack'. If the tradition is right that such boats date in Europe from 1712 then it may well be that the Chinese bulkhead principle was introduced twice, first for small coastal fishing-boats at the end of the seventeenth century, and then for large ships a century later.

A NATURAL HISTORY OF CHINESE SHIPS

'There are those who affirm', said the Dominican friar Domingo de Navarette in 1669, 'that there are more Vessels in China than in all the rest of the known World. This will seem incredible to many Europeans, but I, who have not seen the eighth part of the Vessels in China, and have travel'd a great part of the World, do look upon it as most certain.' All the early travellers to China remarked on the abundance of shipping, from Marco Polo onwards (Fig. 188). It was natural, therefore, that a great variety of ships should have developed, and modern research has done a good deal towards classifying and examining them. Here we cannot give an extended review of its achievements, but a few notes may be made about the literature, and about some of the more remarkable vessels mentioned.

Fig. 188. A photograph of some of the salt transport boats awaiting cargo at Tzu-liu-ching [Ziliujing] in Szechuan [Sichuan], illustrating Marco Polo's remarks about the extraordinary abundance of Chinese shipping. The temple on the projection of the city wall, with the 'half-timbered' houses, are very characteristically Szechuanese [Sichuanese].

Systematic lists of shipping types occur on several occasions in Chinese literature throughout the ages. One of the oldest descriptions of classes of warships which has survived dates from AD 759; it was written by a Taoist [Daoist] military and naval encyclopaedist in the Thang [Tang], and we shall have occasion to refer to it when we come to ship armour (p. 259). Sung [Song] descriptions of types also exist, and for the Ming period, in the late sixteenth century, there is also a detailed description of nine warships as well as a classification of the suitability of different types for the defence of various parts of the coast.

Sung Ying-Hsing [Song Ying-Xing], at the conclusion of the passages quoted above (p. 80 ff), went on to describe nine other types of ship, in the course of which he mentioned the 'yuloh' (a self-feathering propulsion oar), the cloth sails of the Chhien-thang [Qiantang] River, and the two-legged masts of the river boats of Kuangtung [Guangdong] province. These occur only in the south. Then there are two Chinese manuscripts of about 1700 that give drawings and paintings of some 84 vessels, and there are some others. The most extensive collection of descriptions is that compiled by Worcester who analysed no less than 243 vessels. In addition there are Western paintings and photographs of Chinese vessels and a collection of model Chinese ships, yet it is also clear that too little study has been devoted to the Chinese traditions of nautical illustration. Here the centrepiece is the *Wu Ching Tsung Yao* [*Wu Jing Zong Yao*] (Collection of the most important Military Techniques) of 1044, in which Tsêng Kung-Liang [Zeng Gong-Liang] carefully describes six types of warship. His text is closely based on a much earlier set of descriptions

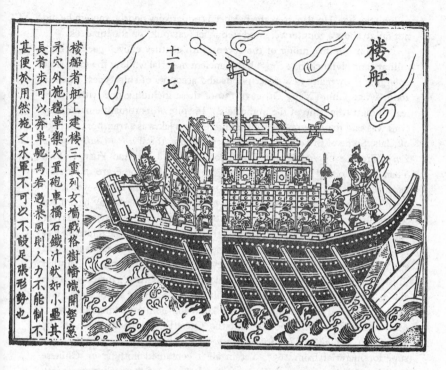

楼舡

十一七

楼船者舟上建楼三重列女牆戰格樹幡幟開弩窗
矛穴外施氊革禦火置砲車檑石鐵汁狀如小壘其
長者步可以奔車馳馬若遇暴風則人力不能制不
甚便於用然施之水軍不可以不設足張形勢也

Fig. 189. A Chinese 'Tower-Ship', from the 1510 edition of the *Collection of the most important Military Techniques (compiled by Imperial Order)*, 1044.

compiled by Li Chhüan [Li Quan] in 759 during the Thang [Tanng]. The six types are:

   (1) 'Tower-Ships' (battleships with fortified upper-works).
   (2) 'Combat-Junks' (less protected).
   (3) 'Sea-Hawk Ships' (converted merchantmen).
   (4) 'Covered Swoopers' (fast 'destroyers').
   (5) 'Flying Barques' (smaller fast ships).
   (6) 'Patrol Boats'.

Unfortunately no corresponding Thang pictures have survived and the *Collection of the Most Important Military Techniques* is illustrated by two series obviously drawn at much later periods. However, those of the 1510 edition may well be contemporary (i.e. Ming), and so are of considerable value. We reproduce one of these – the 'Tower-Ship' – in Fig. 189. The battlements of the upper works at various levels can clearly be seen, and the artist has become so enamoured of the warlike aspect of the ship that he omitted the masts and sails altogether. He also drew the balanced rudder very poorly, though he did not fail to show a counterweighted trebuchet (catapult) on the top deck.

From the beginning of the seventeenth century encyclopaedias both military and civil began to copy all this ancient material while at the same time enlarging it by many new descriptions and accounts of ship types. Thus in 1609, Wang Chhi [Wang Qi] gave 130 of these including one with a two-legged mast (the 'ship of the immortals'). In spite of its romantic name, this craft was said to ply upon the southern rivers and lakes as a transport for high officials. Some of these items were incorporated in the *Wu Pei Chih* [*Wu Bei Zhi*] (Treatise on Armament Technology) of 1628 by Mao Yuan-I [Mao Yuan-Yi]. Most of the drawings here are crude, but some are of interest because they show craft with what seem to be rudders at the bow as well as the stern (the 'eagle-ship' and the 'double-headed ship'). There may have been confusion here with the long bow sweep tillers which certainly existed on river ships for many centuries; but it seems rather more likely that the reference was to types of craft (now mainly Indo-Chinese) which have a rudder at the stern and a centre-board very far forward or even sliding on the stempost. Another illustration shows two barges made fast alongside supporting a turret or turrets above (the 'mandarin-duck rowed assault boat'), and there are of course also rafts with explosives, fire-ships and the like. Wang Chhi's drawings and two of Mao Yuan-I's were incorporated into the 'Great Encyclopaedia' of 1726, the *Thu Shu Chi Chhêng* [*Tu Shu Ji Cheng*], though this contained also some superior material from 1637. Yet in all it contained nothing on Chinese shipping that had not already been given in books of the early seventeenth century.

The analysis of these materials (and they are far from exhausting the available Chinese literature) has not been particularly happy. Some scholars have been unaware of the work of the others, added to which is the inability or unwillingness of Chinese book-illustrators in olden days to depict machinery accurately. Thus Western readers have received yet one more crass misconception of Chinese technology; for what may be laudable in the eighth century becomes scandalous in the eighteenth – in 'unchanging' China quite as much as everywhere else. Moreover, the use of the word 'encyclopaedia' for translating the title 'Great Encyclopaedia' is very unsatisfactory, since it invites comparison with modern encyclopaedias such as the *Encyclopaedia Britannica*. Yet the compilers of the 'Great Encyclopaedia' did not set out to produce a work of reference giving the most up-to-date information on all subjects, but rather to compile a vast anthology of all the classical historical and literary remains. One must remember the proper meaning of the Chinese title: 'Imperially Commissioned Compendium of Literature and Illustrations, Ancient and Modern'. To suggest that the ship design in this book represented either the nautical conceptions or the actual practice of the Chinese of 1726 (or even indeed of 1426) was only to give Westerners still further historically unwarranted grounds for their customary self-satisfaction. In fact an ocean of further research stretches out before us.

Geographical factors have had a considerable influence in differentiating the craft found along the coasts of China. This was clearly seen already in the seventeenth and eighteenth centuries by acute observers of local customs.

In 1673, the scholar Hsieh Chan-Jen [Xie Zhan-Ren] commenting on a passage in the *Jih Chih Lu* [*Ri Chih Lu*] (Daily Additions to Knowledge) of Ku Yen-Wu [Gu Yan-Wu], itself finished the same year, wrote:

> The sea-going vessels of Chiang-nan [Jiangnan] are named 'sand-ships' for as their bottoms are flat and broad they can sail over shoals and moor near sandbanks, frequenting sandy (or muddy) creeks and havens without getting stuck. . . . Chekiang [Qiantang] ships . . . (are built in the same way) and can also sail among sandbanks, but they avoid shallow waters as they are heavier than the sand-ships. But the sea-going vessels of Fukien [Fujian] and Kuangtung [Guangdong] have round bottoms and high decks. At the base of their hulls there are large beams of wood in three sections called 'dragon-spines'. If (these ships) should encounter shallow sandy (water) the dragon-spine may get stuck in the sand, and if wind and tide are not favourable there may be danger in pulling them out. But in sailing to the South Sea (Nanyang) where there are many islands and rocks in the water, ships with dragon-spines can turn more easily to avoid them.

Here the reference to the better sailing qualities of ships with deep hulls and centre-boards is rather clear. With this passage in mind we can look again at Fig. 184, where the *lung-ku* [*long-gu*] is the central strengthening member of the hull of a Fukienese or Cantonese sea-going junk, with rounded bottom and high decks. Such a timber is still called a *lung-ku* by Chinese shipwrights, but it should not be regarded as a keel in the European sense (to which they sometimes apply the same term), for it is not the main lengthwise component of the vessel, this function devolving rather on three or more enormous hardwood wales which are built into the hull at or below the water-line. The real value of the passage is that it points up the differences in hull shape induced historically by the geographical differences between the northern and southern parts of the Chinese culture-area.

North of Hangchow [Hangzhou] Bay the coastal and sea-going craft are flat-bottomed and have a pronounced ridge with relatively large, heavy and square rudders which can be lowered well below the ship's bottom or raised up high. They are thus fitted for frequent beaching in the shallow harbours or muddy estuaries of the north, where the tidal effects are most noticeable, while at sea the rudder acts as an efficient 'drop-keel'. South of Hangchow Bay the coastal waters are deeper, the inlets fjord-like, and the islands more numerous. Here the underwater lines of the vessels become progressively more curved, with a sharper entry, less pronounced ridge and rounder stern; at the same time the rudders, often supplemented by centre-boards, become sometimes narrower and deeper, sometimes drilled with holes and shaped like a rhomboid.

As for the flat bottom and the rectangular shape of the ship in cross-section, it is certainly interesting that this build became generally adopted throughout the world for iron steamships in modern times. Among medieval shipbuilders in wood, the Chinese alone had it. But as we have seen, Chinese ships were not always flat bottomed; though lacking any true keel, their sides sometimes rose up in a quite rounded way from the lowest lengthwise timbers. This we see from texts of a much earlier date, such as the *Kao-Li Thu Ching* [*Gao-Li Tu Jing*] (Illustrated Record of an Embassy to Korea) of 1124. Speaking of the Fukien-built [Fujian-built] 'Retainer Ships', which carried the ambassador's staff, and were somewhat smaller than his own 'Sacred Ship', Hsü Ching [Xu Jing] says:

> The upper parts of the ship (the deck) are level and horizontal, while the lower parts sheer obliquely like the blade of a knife; this is valued because it can break through the waves in sailing.... When the ship is at sea, the sailors are not afraid of the great depth of water, but rather of shoals, for since the bottom of the vessel is not flat, she would heel over if she went aground on an ebb tide. For this reason they always use a lead weight on a long rope to take soundings.

Such a shape of hull could be seen in modern times in certain types of Chinese construction, as in the fishing-boats of Chusan [Zhusan] called 'pairers' and the smaller naval junks used towards the end of the Chhing [Qing] dynasty. And all sea-going junks of the south have it.

Among the most extraordinary of all Chinese craft are the crooked-bow and crooked-stern boats of the upper Yangtze [Yangzi] region. The crooked-stern junks (*wai-phi-ku* [*wai-pi-gu*]) centre on Fouchow [Fuzhou], east of Chungking [Chongqing], at the mouth of the Kungthan [Gongtan] River; but the crooked-bow boats are used further west, for bringing the salt down along an almost un-navigable river from Tzu-liu-ching [Ziliujing]. In both these cases the 'rectangular' bows or sterns are slewed round in such a way as to bring one of their corners more or less in line with the main axis of the vessel. This is done at Fouchow by bending the bottom planks under heat and steam along a line making an angle of some 60° with their long axis, not at right angles. The end bulkheads, moreover, are not quite vertical. The final result is as shown in Fig. 190.

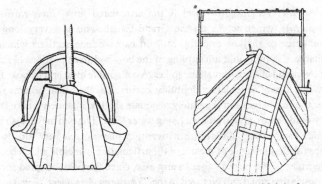

Fig. 190. Drawings of twisted bow (left) and twisted stern (right) of some boats on the Upper Yangtze [Yangzi]. From G.R.G. Worcester.

Fig. 191. Model of boat with a twisted stern. In the Maze Collection, Science Museum, London.

In all these types of boat, very large stern sweep rudders are carried, and in the Fouchow [Fuzhou] crooked-stern boats there are two, which owing to the construction, do not interfere with one another (Fig. 191). Whether the balance of the whole vessel is so affected by the unsymmetrical lines as to confer real advantages in negotiating rapids has not been scientifically investigated, but the boatmen maintain this, and there is at any rate no reason for denying that this unique build is very ancient. The most obvious explanation would be that at some time unkown, it was desired in these parts of China, to obtain a pivot-point for the stern-sweep oars firmer and more closely linked with the whole structure of the boat, than could be obtained by mounting it centrally in the usual way. Such a view would not contradict an invention of the axial stern-post rudder even earlier, since this would not necessarily have penetrated to these western provinces.

Another remarkable vessel is the articulated junk (*liang-chieh-thou* [*liang-jie-tu*]), which works on the Grand Canal. This is a very long and narrow barge of shallow draught, built in two separate sections which are detachable. Coupling and uncoupling of the bow and stern portions is simply done, by the aid of a hook and handspikes. A collapsible mast and leeboards are fitted. The idea of articulation doubtless occurred as the Canal became silted up, for the two halves can separately negotiate shallow channels where larger boats would be forced to wait for a rising water level. The two halves also bank in beside each other. Though this invention may not be very old, it was perhaps the first instance of the use of the articulation principle which became so important in the railway age. In any case, since the boat, adapted for war purposes, is described and pictured in the *Treatise on Armament Technology* of 1628 by Mao Yuan-I [Mao Yuan-Yi], the late sixteenth century would seem to be a likely time for its origin. Figure 192 shows the vessel in this form; it is taken from a manuscript of a related work by the same author, the *Wu Pei Chih Shêng Chih* [*Wu Bei Zhi Sheng Zhi*] (The Best Designs in Armament Technology). Here we see the forward section filled with bombs and other explosives, while the crew and marines occupy the rear section. Presumably the idea was to work the vessel up some creek on a dark night to a point beside some city wall or under a bridge, then uncouple and silently withdraw, having set a time fuse.

Affinities and hybrids

It may be that those who have denied any connection between Chinese and ancient Egyptian shipping have been rather too hasty. The most typical and characteristic Egyptian ship was that familiar type which has an extremely long bow and stern sloping and tapering in each direction above the water-line. The actual length that floats may not be more than half the ship's total length. But this was not the only kind of hull known in ancient Egypt. Especially during the VIth dynasty (about 2450 BC) a quite different type was prevalent, and it presents some extraordinary similarities with certain river junks still in use in China today. Egyptologists know these two types as the Naqadian and Horian, for the first can be traced back to the pre-dynastic pottery designs of Naqad in Upper Egypt, while the second is associated with a conquering people who came from further east and worshipped the god Horus (Fig. 193). One has only to compare the second type, with its high stern, stern gallery, low bow, relatively shortened ends, and mast set forward of the centre-line, with existing Chinese vessels, to see the similarity.

Both types of ship are sometimes seen in one and the same carving, and fortunately tomb-models of both kinds have survived and been examined from the point of view of naval archaeology. A point of great significance emerges, namely that some of the Horian boats are square-ended, like all true Chinese

Fig. 192. An articulated barge of the late sixteenth century in use for military purposes; from the *Wu Pei Chih Shêng* [*Wu Bei Zhi Sheng*] (The Best Designs in Armament Technology).

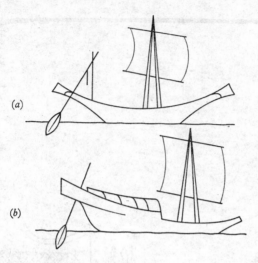

Fig. 193. The two principal types of ancient Egyptian ship: (*a*) Naqadian; (*b*) Horian.

ships. However, as far as can be ascertained (for many of the models are solid blocks), the use of the bulkhead was unknown; at any rate when the model is hollow there are only a number of arched flat timbers across it. However, the gallery at the stern is as characteristically Chinese as the transom stem and stern. One Egyptologist has suggested that the Horian people came from Mesopotamia; in this case certain elements of Chinese shipbuilding might join with other proto-scientific material which, in astronomy for instance, we have seen (pp. 83ff., volume 2 of this abridgement) may have been diffused from Babylon in Mesopotamia to the very early Chinese civilisation. It has been claimed that the Horian ship died out after 2000 BC, but of course it may be safer to say that after that time it appeared no more in carvings or tomb-models.

The two-legged masts of southern Chinese shipbuilding (see Fig. 194) have already been mentioned. Certainly such masts were quite characteristic of ancient Egyptian ships, and there really seems no reason for rejecting this trait as a possible transmission eastwards in very early times. Yet if this should have happened, it is curious that it did not spread further north in the Chinese culture-area. It is clear, of course, that from an engineering point of view such masts are excellent, and have come back with the use of metal tubing for mast construction. Moreover, two-legged masts are recommended by modern yacht designers to avoid interference with the airflow over the leading edge of the sail. And in the days when wood was used, two lesser spars could substitute

Fig. 194. River-junks starting downstream from Kweilin [Gueilin] on the Li-chiang [Lijiang] River in the early morning mists. The two-legged masts so typical of this region recall those of ancient Egypt. Photograph Groff and Lau.

for a more massive one if that was not available. The greater practical value of such a simple invention may have made it more likely that it was independently discovered in both East and West, though we must ask ourselves how simple an invention it really was. There is also the point that diffusion may have been more likely when a practice was magical rather than practical, as in the case of the oculus – the painted eye on the bows – which must certainly have spread outwards in all directions from ancient Egypt or Mesopotamia. When it reached China we do not know, but the fact that it is confined there to the southern and central regions suggests that it came relatively late, perhaps not before the Han.

Several general questions may now be asked. Were there any exceptions in the Chinese culture-area to the basic principle of boats without keel, stem or stern posts? The answer to this is that there was one, and it is of great interest. The dragon-boats (Lung chhuan [Long chuan]), used for the races (Fig. 195) which form such an important feature of the Fifth Month festival, supposed to commemorate the poet Chhü Yuan [Qu Yuan] (died 288 BC), are built with a true keel. This takes the form of a fir pole the length of the boat, which may be as much as 35 metres and resembles an English 'eight', though longer and narrower. It may contain 36 or more paddlers. Though bulkheads slotted to the keel are built in, we are here clearly in the presence of an archaic element of one of the constituent cultures which fused to give the Chinese civilisation. The culture concerned was the south-eastern Malayo-Indonesian one. And there is a further point of particular concern. In order to prevent the bow and stern of so long a vessel from drooping, a long bamboo cable is strung from the projecting bow end of the keel to the stern, exactly as was common practice in ancient Egyptian ships. This again is perhaps an early transmission eastwards,

Fig. 195. Dragon-boat races at Hongkong.

and one is hardly surprised to find it in so archaic a form. The dragon-boat might indeed be described as a vessel derived from the canoe that has survived alone in a world of boats whose ancestor was the raft.

Another possible exception is found in the boats of the Erh Hai in southwestern Yunnan. These seem to have frames without bulkheads, but the presence of keels needs further investigation. The area is a very isolated one, and could well have been influenced by India.

To what extent did regular construction spread from China to other lands? Information here is scanty, yet so far as may be ascertained, all Japanese hulls before the seventeenth century followed approximately the build of the Chinese shipwrights. Later vessels were often copies of western models, notably Russian schooners.

A somewhat more interesting, though much more modest, problem concerns the origin of the punt as we know it today in Western countries. This familiar rectangular craft is very similar to the Chinese sampan, and it is curious that no one has investigated its origin. The word 'punt' is, of course, related to the Latin word for a bridge. There were 'pontones' in Roman times, but apparently they were merchantmen. The oldest illustrations of punts seem to be those in a manuscript of the Emperor Maximilian I (1458 to 1519) and preserved at Vienna. Yet about 630 the scholar Isidore of Seville described the 'ponto' as a simple rowed boat of rectangular shape with sloping sides and a flat bottom; and the *Pandects* (a digest by the Roman Justinian of the writings of Roman jurists) dating from 533 refer to it being used for ferry services.

The problems of the punt's affinities and transmission are therefore rather obscure. However, punts are used widely in Kashmir and the Upper Indus Valley, and if the Chinese build had already found its way westwards in Bactrian times (between 600 BC and AD 600) when Bactria itself (the area which is now part Afghanistan and part Uzbekistan) was a noted meeting point of overland trade routes, punts in our sense may well have existed in Roman Europe. On the other hand if the Roman boats were not what we should call punts, those of St. Isidore may have been introduced through Byzantine contacts. Or perhaps Chinese technicians following the Mongol armies constructed rectangular keel-less boats for military purposes in thirteenth-century Eastern Europe – indeed the punt, like the crossbow, may have been introduced several times. Or lastly the European punt may have always existed as a descendant of the square-ended boats of ancient Egypt just described.

Did hybrid types develop in south-eastern Asia in the regions where the Chinese mingled with Indian and local influences? Abundantly, so research has shown. A keel was the first feature to be adopted, then bulkheads were abandoned for rib frames, as in the *twaqo* of Singapore. Cases are known, such as the Malaysian *tongkung* in which the ships are designed, built, owned and worked by Chinese, yet the structure and rig is entirely European. The Chinese

sails, however, on account of their great efficiency, are among the last components to disappear, and indeed, as in the famous Portuguese 'lorcha' of Macao
and Hongkong, have co-existed since the sixteenth century with a slender hull
of European type.

NAUTICAL HISTORY

The Chinese ship in language and archaeology

If it is not putting too much weight on what might have been only an
arbitrary convention adopted by scribes, the Shang and Chou [Zhou] forms of
words for boats indicate that already in remote antiquity the characteristic
build of the Chinese vessel had developed. They seem to give a remarkable
pictorial statement of the basically rectangular shape, though a design with
pointed ends would hardly have been too difficult to draw. The earliest form of
the word *chou* [*zhou*] 舟 as seen on oracle bones shows the crosspiece and
bulkhead construction, without any sharp stem or stern – or, at least, a
rectangular slightly concave raft with pieces across it. The original significance
of the other component in *chhuan* [*chuan*] 船, meaning a ship, is not known, but
the mouth probably represented the crew, and the two lines river-banks.

From antiquity to the Han

There seems no reason to doubt the historical truth of the fleet which
was sent northwards by the state of Wu, under Admiral Hsü Chhêng [Xu
Cheng], to attack the state of Chhi [Qi] in 486 BC. What he probably commanded was a number of large paddled canoes, some perhaps big enough to
carry deck-castles for archers, and they certainly kept close in-shore. There
was also the great fleet of sailing-rafts built by the King of Yüeh [Yue], another
southern state, in 472 BC. The vessels of the Warring States period, however,
were not all naval, and we can be sure that there were trading expeditions at
least along the coasts of Siberia, Korea and Indo-China (i.e. Vietnam, Laos
and Cambodia). There were also some explorations of the Pacific itself. And,
of course, as ever, inland water transport.

For the Han more information is available than for ancient times, and
only a few points can be touched on here. In 219 BC, the emperor Chhin Shih
Huang Ti [Qin Shi Huang Di] had sent a great military expedition to conquer
the southern peoples of Yüeh [Yue]; its main strength consisted of 'marines'
based upon war-boats with deck-castles. A century later in 112 BC when the
province showed signs of becoming independent, Han Wu Ti [Han Wu Di],
had to send another expeditionary force. This again employed a fleet of 'the
ships of the south which have deck-castles (or, more than one deck)'. Its
commanders were known as admirals, indicating perhaps the growing importance of naval techniques. Indeed, one such title – the 'Torrent–Descending

Commander' – gives a useful glimpse of a kind of seamanship important at the time. Another force went the next year to eastern Yüeh, then in 108 BC Yang Phu [Yang Pu] led a sea force against Korea. Thus naval operations were conducted on a considerable scale during the reign of Han Wu Ti.

At the beginning of the Later Han period, in AD 33, a remarkable floating bridge and boom in Hupei [Hubei] was destroyed by a Han fleet of thousands of vessels including many 'castled ships'. Just ten years later there was a great expedition to Tongking involving 2000 'castled ships', (*lou-chhuan* [*lou-chan*]), and subsequently many naval fights between the Chinese and the Champa people (modern Annam) are recorded. The term *lou-chhan* for large warships persisted down through the centuries. By the eighth century, the best Thang [Tang] source tells us that they had three decks, with bulwarks, arms, flags, and catapults, but were not very handy in rough weather. There is some evidence that they engaged in ramming, like Greek triremes.

From the beginning of the Han era, sea communications with Kuangtung [Guangdong], Indo-China and Malaya began to become important. In AD 2 Wang Mang got tribute of a live rhinoceros from those parts, and it certainly came some of the way by boat. This tribute was repeated in AD 84 and 94, and continued intermittently as late as the Thang [Tang]. An interesting passage in the *Chhien Han Shu* [*Qian Han Shu*] (History of the Former Han Dynasty) describes Han trade with the South Sea:

> There are superintendent interpreters belonging to the civil personnel, who recruit crews and go to sea to trade for brilliant pearls, glass, strange gems and other exotic products, giving in exchange gold and various silks. In the countries where they come the officials and their followers are provided with food and handmaidens. Merchant ships of the barbarians (may) transport them (part of the way) home again. But (these barbarians) also, to get more profit (sometimes) rob people and kill them. Moreover (the travellers) may encounter storms and so drown. Even if nothing (of this kind happens, they are) away for several years.

This probably refers to the two centuries preceding the time when Pan Ku [Ban Gu] was writing the history (i.e. about AD 90), so we may take it as well applicable to the first century BC, indeed back to the time of Han Wu Ti [Han Wu Di]. Since the account speaks of voyages taking just over a year, and yet lacks any legendary quality, it seems likely that the missions went to the western boundaries of the Indian Ocean; archaeological evidence confirms this.

Indeed, it is more than likely that the foundations of maritime trade had already been built by the people of Yüeh [Yue] in the warring States period. A passage in the *Chuang Tzu* [*Zhuang Zi*] (The Book of Master Zhuang), seemingly often misunderstood, may be brought forward in witness of this. The Taoist [Daoist] recluse Hsü Wu-Kuei [Xu Wu-Gui], having had an interview with Duke Wu of Wei, is discussing his good reception with the Duke's minister:

> Have you not heard [he says], of the wanderers of Yüeh? When they have been gone from their country several days, they are glad when they see anyone whom they knew there. When they have been absent for weeks or months, they are happy if they meet anyone whom they had formerly seen at home. But by the time they have been away for a whole year they are delighted if they meet with anyone who even looks like a compatriot. The longer they are gone, the more affectionately they think of their own people – is this not so?

Thus the Duke had wandered far from his native land so no wonder he welcomed a messenger from home. But for us the interest lies in the sails of the merchantmen of Yüeh flitting about the isles of the Indies.

It may not have been only the Indies. The possibility is still open that Han trading envoys got as far as Ethiopia, at least according to one plausible interpretation of places mentioned and the timings given by Pan Ku [Ban Gu] in his history.

Representations of boats and ships of the Warring States and Han times were until lately scarce. Tomb-shrine reliefs carved between AD 125 and 150 all show sampans carrying two or three people each, but their nature is not clear; they may be dugout canoes or craft made from bundles of reeds. More interesting are the Warring States and Early Han bronzes depicting war-boats which show distinctly above the rowers an upper deck carrying spearmen, 'dagger-axe' halberdiers and archers. This is the first appearance of the 'castled ships' already mentioned.

A sidelight on these ships may be obtained from inscribed bronze drums of the same period found in Indo-China. These show in a stylised form manned superstructures and thus are a rather precious record of the kind of war-boats which the Chinese 'Wave-Subduing Admirals' had at their command.

In the late 1950s a magnificent model of a river-boat more than a metre long was found in a Han tomb that seems to date from about 49 BC. Delicately made of wood, it has sixteen oars and a great stern-sweep oar of twice their length (Fig. 196). Typically Chinese in build, it has a flat bottom, a rectangular bow and stern, while that part of the bow projecting above the water is very elongated. Unfortunately it has no mast and sails. Published photographs show considerable differences in the way the boat can be assembled. In one reconstruction a couple of deck houses rise one above the other, in a second they follow on fore and aft. The former shows a large stern gallery, the ancestor

Fig. 196. Wooden tomb-model of a Former Han river-ship excavated from a princely burial of the first century BC at Chhangsha [Changsha]. Length 1.29 metres. There is some uncertainty as to how the component pieces should be put together. The arrangement shown here is about the same as that adopted at the National Historical Museum at Peking [Beijing] (1964), but to the nautical eye it cannot be right, as it leaves no room either for helmsman or rowers (the black object amidships is the smallest of the three deckhouses). The arrangement by the scholar Hsia Nai [Xia Nai] is preferable. In this the U-shaped piece, here enclosing the after deckhouse, ought to project astern as a gallery, and a central notch in this indicates that the stern-sweep or steering-oar was intended to turn on it, though one would have expected it rather to work through it on the terminal hull transom bar. The two larger deckhouses ought perhaps to be superimposed, and the smallest one should be moved either aft or forward. Lastly the bulwarks are upside down, for the oar ports must be along their lower not their upper edges; and this is proved because those along the line of ports are flat while the other edges are slightly convex. Other components of the tomb-model have not been incorporated in any reassembly seen by Dr Needham. There are no obvious bulkheads.

of those elaborate structures on sea-going junks which cover the rudder housing, but the latter, less plausibly, omits it and leaves no room for the steersman. It is not easy to pick out any existing Chinese craft which resemble this boat in general, but there is something archaic (indeed truly Egyptian) about the long projecting bow and stern in what seems to be a 'Horian' style of boat.

By remarkable good fortune, excavations at Canton since 1954 have given us fine models of craft from the first century AD to complement the river-ship model, since they are of vessels from the estuaries and the open sea. Some are of red pottery and rather roughly finished, though they do show

Fig. 197. Tomb-model of a ship in grey pottery of the Later Han period (first century AD), excavated from a burial under the city of Kuangchow [Guangzhou]. Length just under 56 centimetres. This is a view from the starboard side, showing the slung rudder under the overhanging poop or 'false stern', steersman's cabin, several roofed or matting-covered deckhouses, a long poling gallery, bollards or yuloh thole-pins, and a projecting bow with an anchor hanging from it. The mast was probably stepped just in front of the deckhouses. Photograph by Canton Museum.

clearly the typical square-ended bow and stern and the flat bottom of the hull. However, a later Han tomb has produced a more detailed model in grey pottery (Fig. 197). The hull is of the standard Chinese type, and there is an ornamental structure at the bow which may be the ancestor of the 'prow yokes' (bars across the prow) now mostly found on Indo-Chinese vessels. Passing aft we find a sort of cowl, doubtless made of bamboo matting, to provide protection for the forward well-deck (an open space on the main deck), frequently seen on medieval Chinese pictures of ships, and still in use today. There are three bollards or rowlocks on each side, between which stand two figures amidships. For two-thirds of its length the vessel is covered by a barrel-vaulted roof (perhaps of matting), out of which stand three deck-houses, the one at the stern being clearly the steersman's cabin. The covered part is flanked by three tall pointed timbers whose function is uncertain, and by well-marked poling galleries outboard of the decks. Another crew member stands on the starboard side. Where the mast was fitted remains a mystery, but the most likely place would seem to be the gap just in front of the covered deck where the poling galleries end. It is sad that all evidence of mast and rig is gone, but the general resemblance to a traditional Cantonese ship of a few years ago being worked upstream can be glimpsed from Fig. 198.

The third century, the San Kuo [San Guo] period, is rich in naval history, but unfortunately the annalists were sparing of detail about the ships themselves. Doubtless the tower ships were larger, and we begin to hear of fast fighting ships. Bulwarks were covered with wetted hides to nullify the effect of

Fig. 198. The use of the poling gallery on a Cantonese ship at Chhü-chiang [Qujiang] on the Pei-chiang [Beijiang] River, taken in 1944 by Dr Needham from a passing sampan. This is gruelling labour upstream on windless days.

incendiaries, but sometimes these weapons were very effective, as in the famous case of the Battle of the Red Cliff (AD 207), when Tshao Tshao's [Cao Cao's] fleet was destroyed. Sea-going ships as well as river craft must have been developing, for in 233 a fleet of the state of Wu was lost in a storm in the Yellow Sea. From a number of descriptions it is clear that detachments of crossbowmen (unknown until much later in medieval Europe) firing from the ships often played a more important part than boarding or ramming.

In this period the word *po* [*bo*] for sea-going ships or junks, appears. Chinese diplomats such as Khang Thai [Kang Tai] who travelled to South East Asia, and merchants like Chia Hsiang-Li [Jia Xiang-Li] who traded all the way to and from India, sailed in a *po*. To the Chinese, for whom the building of sharp-ended boats with stem and stern posts was only a comme-morative ritual act, Khang Thai brought knowledge of such craft still in full use in Cambodia. Indeed, such boats, obviously related to the Chinese dragon-boats, continued in service in South East Asia for centuries, so that the late thirteenth-century traveller and writer Chou Ta-Kuan [Zhou Da-Guan] was able to give a description of Cambodian shipbuilding:

The great boats are made of hard wood. The carpenters have no saws and work with hatchets, so that it takes much wood and much trouble to produce a single plank. Knives are also used, even in house building. For the boats they use iron nails, and cover them (i.e. roof them) with *chiao* [*jiao*] leaves, held in place by strips from the *pin-lang* [*bin-lang*] trees (*Areca Catechu*, the betel-nut palm). Such a boat is called a *hsin-na* [*xin-na*] and is rowed. For caulking they use fish oil and lime. Small boats are made of great trees hollowed out in the form of a scoop softened by fire, (and water), and enlarged by ribs, so that they are broad in the centre and pointed at the ends. They have no sail, but can carry several persons, and they row them. They are called *phi-lan* [*pi-lan*] boats.

His description of the dugout and ribs indicates that these were unfamiliar to the Chinese of his time. Another text of the third century is a very important one because it establishes a date for the earliest development of fore-and-aft sailing. We shall come to this later (p. 193).

For the Sui and Thang [Tang] periods (sixth to tenth centuries) information is hard to assemble. But it does seem that it was not, apparently, until the close of this period that Chinese ships grew to their full stature, for we know that about 587, the engineer and high official, Yang Su, was constructing vessels with as many as five decks, and more than 30 metres in height. However, no one seems to have noticed a rather important passage in the *Thang Yü Lin* [*Tang Yu Lin*] (Miscellanea of the Tang Dynasty), compiled in the Sung [Song] from Thang [Tang] documents by Wang Tang [Wang Dang]. It refers to the eighth century:

On the Yangtze [Yangzi] and the Chhien-thang [Qiantang] Rivers they set sail according to the two tides, and Chiangsi [Jiangxi] is the place where shipping flourishes most.

Sails are plaited from rushes, and the largest of them exceeds 80 sections. From Pai-sha [Baisha] (White Sands) the ships go upstream when they have a north-east wind; this is called the 'reliable seasonal wind'. In the seventh and eighth months there is the 'upper reliable (seasonal) wind'; in the third they rely on (migrating) birds and in the fifth they look for (the wind on) the wheat. 'Catapult clouds' presage a storm.

... ...

The sailors have a saying 'Water won't carry 10 000' – this means that the largest ship cannot exceed a load of 8000 to 9000 piculs (some 560 to 635 tonnes).

In the Ta-Li [Da-Li] and Chen-Yuan [Zhen-Luan] reign-periods (766 to 779 and 785 to 804) there were the (large) ships of Yü Ta-

Niang [Yu Da-Niang]. The crews of the ships lived on board, they were born, married and died there. The ships had, as it were, lanes (between the dwellings), and even gardens (on board). Each one had several hundred sailors. South to Chiangsi [Jiangxi] and north to Huainan they made one journey in each direction every year, with great profit. This almost amounted to 'carrying 10 000'.

In Hupei [Hubei] many people live entirely on the water, in boats nearly half as big as houses. The large boats are always owned by merchants, who have bands of musicians on them, as well as their slave-girls, and all these people live under the poop.

The sea-going junks are foreign ships. Every year they come to Canton and An-i [Anyi]. Those from Ceylon are the largest, the companion-ways alone being many metres high. Everywhere the various kinds of merchandise are stacked up. Whenever these ships arrive, crowds come forth into the streets, and the whole city is full of noise.... When the ships go to sea, they take with them white (homing) pigeons, so that in the case of shipwreck the birds can return with messages.

This passage attests the use of mat-and-batten sails and the existence of remarkably large river-junks working back and forth between Chiangsi and Anhui; it also strongly indicates that the axial rudder was in use. It seems to suggest, too, that these vessels had not been entirely successful, and had been replaced in the writer's own time by rather smaller ships. It was about this period that the Arab, Persian and Sinhalese merchants came on further north than Yangchow [Yangzhou], while the Shantung [Shandong] ports and the old estuary of the Huai River were the resort of Japanese and Koreans. From what Wang-Tang says we must not infer that all the coastal and sea-going ships were foreign, since from the beginning of the eighth century it became the practice to carry large amounts of grain and other commodities from the south to Hopei [Hobei], the northern province menaced by Chhi-tan [Qi-dan] and the Koreans. It was this period that saw the peak of maritime intercourse between China, Japan and Korea.

There are some beautiful paintings of small sailing-boats preserved in Chinese frescoes at the cave-temples at Tunhuang [Dunhuang] in Kansu [Gansu] province. They occur in all kinds of contexts. For example, there is the Buddhist Ship of Faith, whose square-ended bow and stern are typically Chinese of the early Thang [Tang] (Fig. 231), and also with a characteristic rear extension of the uppermost lengthwise timbers. But here the Chinese elements end – except for the beehive hut of straw which someone thought would be the right thing for a deck-house. The bellying-square sail is exceedingly un-Chinese, though suitable for a boat of the Ganges. Perhaps the monastic artist, who had quite possibly travelled overland from India and never seen either the sea or the great rivers of China, was remembering the ships of Bengal rather than depicting those of his new country. All the Tunhuang pictures follow the same pattern, reminding us of the strange hybrid ship painted on the wall of one of the Ajanta caves in western India, which seems to contain elements of Roman, Egyptian, and especially Chinese, design. But the most seamanlike of the Buddhist ships is sculptured on a relief of the style of the Liang period (sixth century). If they are of truly Chinese craft, such representations as these might seem to contradict other evidence which points, as we shall see later, to the antiquity of the mat-and-batten sail. But, of course, it may be that loose square-sails co-existed with the mat-and-batten type for centuries; after all loose square sails are not unknown on Chinese craft free from external influences. Sung [Song] paintings still depict such sails occasionally, but as time goes on they give way to the characteristic flat lug sail.

Numerous sculptures of ships appear on the walls of the great monument at Borobodur in Java. One of these differs from the others, and it shows the rigid Chinese mat-and-batten sail very clearly. As the most probable date for these carvings is AD 800, this may be the oldest depiction of a Chinese seagoing ship. But taking all these representations together, it looks as if there was certainly some mutual influence between Chinese and South East Asian shipbuilding methods during the eighth century.

From the Thang [Tang] to the Yuan

Specific research would certainly discover much of interest for nautical technology between the eighth and twelfth centuries, but as yet only a few notes can be given.

There was great activity in canal and river boat construction about 770 associated with the official Liu Yen, who set up ten shipwright's yards and offered competitive awards. Many-decked naval vessels were used in fighting during 934, and the first Sung [Song] emperor, Chao Khuang-yin [Zhao Kuang-yin] attached importance to shipbuilding and often visited the yards. Then in 1048 the Liao State, conscious also of sea- (or rather river-) power, commissioned the building of 130 warships which would carry horses below decks and soldiers above; these worked effectively as landing-craft in operations along the Yellow River. In 1124 two very large vessels were built for the embassy to Korea, and we read of the excitement of the people there when they arrived in port. In 1170 a traveller on the Yangtze [Yangzi] watched naval manoeuvres carried out by 700 ships each about 30 metres long, with castles, towers, flags flying and drums beating. They sailed rapidly even against the stream.

From the twelfth century there happens to be a cluster of important documents, both pictorial and literary. Let us start with one far away from

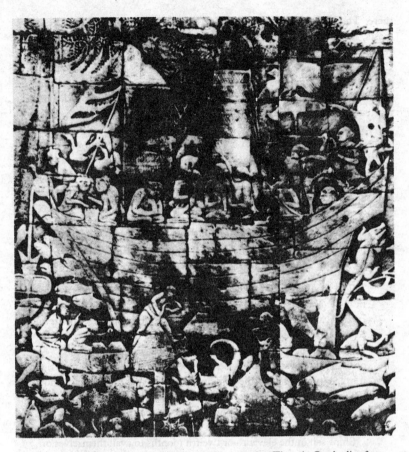

Fig. 199. Carving on the Bayon temple at Angkor Thom in Cambodia of a Chinese merchant-ship; executed about 1185. The mat-and-batten sails, with their multiple sheets, the axial rudder slung below the level of the ship's bottom, the anchor with its winch, and the characteristic 'oriflamme' flag, are all to be noted. Many other vessels are depicted on the monument, but they are invariably of the paddled canoe type, even if substantial in size.

China proper, a wonderful carving in relief on the Bayon temple at Angkor Thom in Cambodia (Fig. 199). Dating from about 1185, the ship here is quite different from any other craft depicted on the monument, which are paddled canoe-like boats similar to those already described by Khang Thai [Kang Tai] (p. 107); it is generally agreed to represent a Chinese junk from Kuangtung [Guangdong] or Tongking. The strakes of the carvel-built hull are clearly shown, and the reduced cross-piece bow typical of the South China junk is there. The stern gallery overhangs in truly Chinese fashion, there are two masts, each carrying Chinese matting sails, and even the multiple ropes that set the sails are shown. The flagstaffs at bow and poop bear flags of typical Chinese design.

A detailed comparison has been made between the Bayon junk and craft still sailing today, built and worked by the Chinese in Siam. It shows that most particulars are the same. Even the deck-houses are in the same position. However, the hull is now hybrid, bulkheads co-existing with a keel and true stem- and stern-posts, though it is suggested that the Bayon ship was already hybrid in this respect. Nevertheless, the Sino-Siamese junks of today carry sails very much longer than those shown on the carving, and with rounded edges. On the whole, then, it seems far more likely that the Bayon ship was a true Chinese vessel in all respects, and not a hybrid at that early date.

Some sixty years before the Bayon junk was being carved, Chu Yü [Zhu Yu], whose father had been Superintendent of Merchant Shipping at Canton and whom we have already met in connection with the history of the magnetic compass (p. 27), wrote of things as they were from about 1086 onwards:

> The Pavilion of the Inspector of Foreign Trade is by the waterside near the Hai-Shan Tower, facing the Five Islands. Below this, the river is called the 'Little Sea'. In mid-stream for some three metres or so the merchant ships take on water for use on their voyages; this water does not spoil, but water from outside this limit, and all ordinary well-water, cannot be stored (on board ship), for after a time it breeds worms. What the principle is underlying this I do not know. [Probably the sailors chose brackish water from bottom springs, not too salt to drink, but salt enough to prevent the growth of algae and animal life.]
>
> Ships sail in the eleventh or twelfth months to avail themselves of the north wind (the north-east monsoon), and return in the fifth or sixth months using the south wind (the south-west monsoon).
>
> The ships are built squarely like rectangular wooden grain-measures.

> If there is no wind, they cannot move. Their masts are firmly stepped, and the sails are hoisted beside them. One side of the sail is close to the mast (round which it moves) like a door on its hinges. The sails are made of matting.
>
> These ships are called *chia-thu* [*chai tu*] – a local expression.

At sea they can use not only wind from the stern, but winds from on-shore and off-shore can also be used. It is only a wind (directly) contrary which can not be used. This is called 'using the winds of the three directions'. When the wind is dead ahead they cast anchor and stop....

According to the government regulations for sea-going ships, the larger ones can carry several hundred men, and the smaller ones have more than a hundred on board....

Sea-going ships are several tens of fathoms in breadth and depth.

The greater part of the cargo consists of pottery, the small pieces packed in the larger, till there is not a crevice left.

At sea (the mariners) are not afraid of wind and waves, but of running aground, for if this happens there is no way of getting off again. If the ship suddenly springs a leak, they cannot mend it from the inside, but they order their foreign blackamoor slaves to take chisels and oakum and mend it from outside, for these men are expert swimmers and do not close their eyes when under water....

This passage, the literary counterpart of the Bayon junk, is of much interest. It provides evidence that at the end of the eleventh and beginning of the twelfth centuries, a bulkhead build was in use, there was a fore-and-aft square lugsail (not in Western Europe until about 1500), and the ships had taut mat-sails to make way up wind. This is all confirmed in other books of the period.

There is also a document of high importance relating to the river-ships of China, almost contemporary with the description of Chu Yü [Zhu Yu]. This is the painting 'Going up the River to the Capital at the Spring Festival' by Chang Tsê-Tuan [Zhang Ze-Duan] made a little before 1126 when Khaifêng [Kaifeng], the capital in question, fell to the Chin [Jin] Tartars. Reproduced in Fig. 200, it shows one ship lowering its two-legged mast before passing under a bridge, while others are to be seen loading or unloading along the banks or being tracked upstream. The junks are broadly speaking of two different types, freighters with narrow sterns, and passenger-boats and smaller craft with broad ones, but both are provided with large and prominent slung rudders. These are all balanced – a remarkably advanced technique which will be discussed later. Large stern-sweep and bow-sweep oars and used on two or three of the boats, some worked by as many as eight men. The history of this painting has been much discussed by scholars, and so famous was it that one Yuan emperor wrote a poem on the one he possessed. The oldest surviving copy is conserved in the Imperial Palace Museum, its silk is recognisably Sung [Song], and it bears an inscription which is dated 1186, just sixty years after the original was completed.

One of the difficulties in this part of our study is that lacking systematic

Fig. 200. One of the passenger-carrying river-junks in the painting 'Going up the River to the Capital at the Spring Festival'; a work of Chang Tsê-Tuan [Zhang Ze-Duan] dating from about 1125, and so from the Liao period. The scene is one of the waterways near Khaifêng [Kaifeng], perhaps the Pien [Bian] Canal. Judging by the figures on board, the length of the vessel from stem to stern would be about 20 metres. It is being tracked upstream by five men (out of the frame to the left), the bipod mast being supported by numerous stays. The large rudder, slung and balanced, is especially noteworthy. The poling gallery is in use at the port bow and to starboard while the master and his mates, interrupted at their lunch on the upper deck, are shouting instructions and warnings to the crew of a large boat (out of the frame on the left) which seems likely to collide with another junk with lowered mast which is negotiating a large bridge.

treatment of the subject in Chinese it is necessary to fall back upon the words of a number of persons who gave, at different dates, general descriptions of Chinese nautical technology. We have heard from some Chinese commentators, and must now allow ourselves to be buttonholed by two other 'ancient mariners', Marco Polo and the Muslim traveller Ibn Battutah.

Marco Polo was in China from 1275 to 1292 and the following accounts were set down about 1295, after his return to Italy. On the way out he saw the sewn ships of the Persian Gulf, 'ships of the worst kind' he called them. Yarn or twine from the coconut palm, or its nuts, was used and fish-oil mixed with oakum, instead of pitch, for caulking. These ships had but one mast, one deck, and one rudder. By contrast, his admiration for the Chinese ships, of all kinds, was unbounded. He described the wealth of the cities Yangchow and Chhuchow [Yangzhou and Zhuzhuo] with the 'marvellous great shipping' that frequented them. Of the Yangtze [Yangzi] he said the 'more dear things, and of greater value, go and come by this river, than go by all the rivers of the Christians together, nor by all their seas'. He estimated that the lower reaches of this river were navigated by 15 000 ships, not counting the many rafts of good timber. He was chiefly impressed, however, by the high freight capacity of the river-junks, and by the cables made of bamboo, as they still are today.

His account of the sea-going junks written in connection with his description of Chhüanchow [Quanzhou] is of great interest and something of it must be quoted:

> We shall begin first to tell you about the great ships in which the merchants go and come into Indie through the Indian sea. . . .
> They have only one floor, which is called with us a deck, one for each, and on this deck there are commonly in all the greater number quite 60 little rooms or cabins, and in some, more, and in some, fewer, according as the ships are larger and smaller, where, in each, a merchant can stay comfortably.
> They have one good sweep or helm, which in the vulgar tongue is called a rudder.
> And four masts and four sails, and they often add to them two masts more, which are raised and put away every time they wish, with two sails, according to the state of the weather.
> Some ships, namely those which are larger, have besides quite 13 holds, that is, divisions, on the inside, made with strong planks fitted together, so that if by accident that the ship is staved in any place, namely that it either strikes on a rock, or a whale-fish striking against it in search of food staves it in . . . And then the water entering through the hole runs to the bilge, which never remains occupied with any things. And then the sailors find out where the ship is

staved, and then the hold which answers to the break is emptied into others, for the water cannot pass from one hold to another, so strongly are they shut in; and then they repair the ship there, and put back there the goods which had been taken out.

They are indeed nailed in such a way; for they are all lined, that is, that they have two boards above the other.

And the boards of the ship, inside and outside, are thus fitted together, that is, they are, in the common speech of our sailors, caulked both outside and inside, and they are well nailed inside and outside with iron pins. They are not pitched with pitch, because they have none of it in those regions, but they oil them in such a way as I shall tell you, because they have another thing which seems to them to be better than pitch. For I tell you that they take lime, and hemp chopped small, and they pound it all together, mixed with an oil from a tree. And after they have pounded them well, these three things together, I tell you that it becomes sticky and holds like birdlime. And with this thing they smear their ships, and this is worth quite as much as pitch.

Moreover I tell you that these ships want some 300 sailors, some 200, some 150, some more, some fewer, according as the ships are larger and smaller.

They also carry a much greater burden than ours.

And formerly in time past the ships were larger than they are now at present, because the violence of the sea has broken away the islands in several places that in many places water was not found enough for those ships so great, and so they are now made smaller, but they are so large that they carry quite 5000 baskets of pepper, and some 6000.

Moreover I tell you that they often go with sweeps, that is, with great oars, and four sailors go with each oar.

For thirteenth-century junks, then, Marco Polo gives evidence for cabins (naturally the first thing a travelling merchant would notice), rudder (this had already been in use for eighty years or so in Europe), multiple masts (not yet used in Europe), and bulkhead-built hull. He makes a particular point about the repair of the ship by the continual overlaying with new layers of caulked strakes. This system of superimposed timbers ('doubling') was afterwards employed in European warships of the eighteenth and early nineteenth centuries. The oars rowed by four men each may well have been yulohs, though the description leaves it uncertain. In a passage not quoted here, Marco Polo concerns himself with the way the ships sailed, and from this it would appear that for some reason or other the smaller ships could sail better to windward than the larger, so that under such circumstances they had to resort to towing by the smaller tenders.

When Marco Polo left China after seventeen years, it must have been a hard parting for him, even though he was going as imperial envoy-extraordinary in charge of a princess. 'The Great Khan caused to be armed and set forth fourteen great ships, and every one of them had four masts ... In every ship he put 600 men, and provision for two years ....' It was almost certainly a much braver fleet than any European country, including the England of Edward I, and the France of St Louis, could have launched for the occasion.

While Marco Polo's words are fresh in our minds, let us listen to the description of Ibn Battutah, who was in China just over half a century later (1347). His opening statement is significant:

> People sail on the China seas only in Chinese ships, so let us mention the order observed upon them.
>
> There are three kinds: the greatest is called the 'jonouq', or in the singular, 'jonq' (certainly *chhuan* [*chuan*]); the middling size is 'zaw' (probably *tshao* [*cao*] or *sao*)' and the least a 'kakam'.
>
> A single one of the greater ships carries twelve sails, and the smaller ones only three. The sails of these vessels are made of strips of bamboo, woven into the form of matting. The sailors never lower them (while sailing, but simply) change the direction of them according to whether the wind is blowing from one side or the other. When the ships cast anchor, the sails are left standing in the wind.
>
> Each of these ships is worked by 1000 men, 600 sailors and 400 marines, among whom there are archers and crossbowmen furnished with shields, and men who throw (pots of) naphtha.
>
> Each great vessel is followed by three others, a 'nisfi', a 'thoulthi' and a 'roubi'.
>
> ... ...
>
> The pieces of wood, and those parts of the hull, near the water (-line) serve for the crew to wash and to accomplish their natural necessities.
>
> On the sides of these pieces of wood also the oars are found; they are as big as masts, and are worked by ten to fifteen men (each), who row standing up.
>
> These vessels have four decks, upon which there are cabins and saloons for merchants. Several of these 'misirya' contain cupboards and other conveniences; they have doors which can be locked, and keys for their occupiers. (The merchants) take with them their wives and concubines. It often happens that a man can be in his cabin

without others on board realising it, and they do not see him until the vessel has arrived in some port.

The sailors also have their children in such cabins, and (in some parts of the ship) they sow garden herbs, vegetables, and ginger in wooden tubs.

The commander of such a vessel is a great Emir; when he lands, the archers and the Ethiops [the black slaves mentioned previously] march before him bearing javelins and swords, with drums beating and trumpets blowing. When he arrives at the guesthouse where he is to stay, they set up their lances on each side of the gate, and mount guard throughout his visit.

Among the inhabitants of China are those who own numerous ships, on which they send their agents to foreign places. For nowhere in the world are there to be found people richer than the Chinese.

Ibn Battutah had his own personal experiences with these ships. At an Indian port the unfortunate man embarked, with a number of his concubines, on a junk, but all the suitable cabins had been reserved by Chinese merchants, so the party transferred to a *kakam*; then before he himself went on board, the junk with the presents for the emperor sailed out into a storm and was lost with all hands. The captain of the *kakam* then also left without him, and he never recovered any of the girls or his valuable merchandise. And on the way home he experienced a tempest and is supposed to have come across that fabulous bird of prey, the roc, mentioned too by Marco Polo.

Nevertheless, his account confirms that of Marco Polo in a number of respects, and also complements him most usefully. He gives evidence for the great mat-and-batten square sails, much greater in number than were carried by any European or Arab ship of the time, and their ability to make use of the wind coming from almost any quarter. As to the large oars, these must have been the self-feathering propulsion 'yulohs' as becomes evident from his development of the subject:

This sea (on the way to China) has no wind, waves, or motion, in spite of its great extent; hence each Chinese junk is accompanied by three boats as we have already said. They serve to make it proceed by towing and rowing. Besides, there are in the junk about twenty very great oars, like masts, each of which have about thirty men to work them, standing in two rows facing each other. The oar, like a club, is provided with two strong cords or cables, and one of the two rows of men pulls on the cable and then lets it go, while the others pull on the second cable. And as they work, these rowers raise good voices in a shanty, generally saying "la, 'la, 'la, 'la".

In the meantime, Marco Polo's information (and that of others less well known) had been spreading in Europe. The famous Catalan world-map of 1375 and the world-map of Fra Mauro Camaldolese of 1459 both contain ship illustrations based on their evidence. The eastern portion of the Catalan map shows in the seas three ships which are recognisably large junks, while Fra Mauro's eastern section map has numerous ships, all of them considerably larger than the European vessels drawn elsewhere on the map.

The discovery by Europeans that really large ships with multiple masts had been built and could do useful work was important. Of course, there had been two- and three-masted ships in Hellenistic times (first to third centuries AD), but these did not survive the fall of the Roman Empire. The importance of this discovery has been emphasised by the historian of naval architecture, G.S. Clowes:

> It was the introduction of the three-masted ship with its improved ability to contend with adverse winds, which made possible the great voyages of discovery of the end of the fifteenth century, of Columbus to the West Indies, of Vasco da Gama to India, and of the Cabots to Newfoundland; and it is a curious thought that this great development may really have been due to the introduction into Europe of accounts of the multiple-masted Chinese junks which traded so effectively in the Indian Ocean....

It is a curious fact too that just as the Europeans were struck by the larger size of Chinese ships, so the Chinese were (or had been) under the impression that the ships of the Far West were also larger than their own. In 1178 the historian Chou Chhü-Fei [Zhou Qu-Fei] had written:

> Beyond the Ocean west of the Arab countries there are countless countries more, but Mu-Lan-Phi [Mu-Lan-Pi] is the only one which is visited by the great ships of the Arabs. Its ships are the biggest of all.... A single ship carries a thousand men; on board there are stores of wine and provisions, looms and shuttles (for weaving), and a market-place. If it does not encounter favourable winds it does not get back to port for years.

All this was repeated by Chao Ju-Kuan [Zha Ru-Guan], another historian, in 1225. But there was seemingly a legendary element here, for the accounts go on to mention grains of wheat five centimetres long, melons almost two metres round, and sheep whose fat could be harvested from time to time by surgical operation. It seems then that people at both ends of the Old World thought that the other end had the largest ships, but objectively the Europeans were right in this opinion while the Chinese seem to have been wrong.

Or were they? There may be more than meets the eye in the story of Mu-Lan-Phi [Mu-Lan-Pi]. The usual identification of the place has been Spain, the name deriving from the al-Morabitun dynasty (1061 to 1147). But the contemporary botanist Li Hui-Lin, realising that the time of 100 days seems impossibly long for an east-west Mediterranean transit, suggests that in fact the journey was a trans-atlantic crossing, and the strangeness of the plants and animals described may conceal species typical of the Americas. If one follows him in taking the descriptions seriously the huge cereal grains must be maize, the melons could be the *Cucurbita pepo* which can weigh as much as 100 kg, the sheep llamas and alpacas. Li Hui-Lin associates the idea of Arab trans-atlantic voyages with an old story reported by the Muslim geographer al-Idrisi that in the tenth century some Spanish Muslim sailors set out westwards from Lisbon but were never seen again. But the greatest difficulties in Li Hui-Lin's story lie on the nautical side, for everything known of the sewn ships of Arabic culture stops us accepting that they could ever have been built stoutly enough to withstand a return trans-Atlantic crossing. Moreover, to return to Europe they would have had to discover the system of winds and currents in the Atlantic, which were only laid bare five centuries later by the Portuguese, and we have no evidence whatever that they did so.

Probably none of the foreign travellers in Sung [Song] and Yuan China had sufficient historical perspective to realise that in the Southern Sung a great event had occurred, the creation of the Chinese navy. The development of the maritime south had been the sociological consequence of the wars, invasions, and political unrest, and even of the climatic changes in the north. Masses of the population were driven down to the Fu and Kuang [Guang] coastal provinces with their innumerable rivers, fjords, creeks and havens. Since agriculture supported the people less readily here, commercial cities, actively backed by the state, began to flourish, and this encouraged shipbuilding, navigation, and all else that follows when men, for trade or defence, go down to the sea in ships. Thus it came about naturally that in the first half of the twelfth century, after the fall of Khaifêng [Kaifeng] and the removal of the capital to Hangchow [Hangzhou], when government was centred in the south-eastern quarter of China, the rise of a permanent navy first took place. As the political writer Chang-I [Zhang-Yi] said in 1131, China must now regard the sea and the river as her Great Wall, and substitute warships for watchtowers.

The first Admiralty was set up the following year at Ting-hai [Dinghai] under the name Imperial Commissariat for the Control and Organisation of Coastal Areas. From a total of 11 squadrons and 3000 men, it rose in one century to 20 squadrons totalling 52 000 men, with its main base near Shang-hai. The regular striking force could be supported at need by substantial merchantmen; thus in the campaign of 1161 some 340 ships of this kind took part in the battles on the Yangtze [Yangzi]. The age was one of continual innovation; in 1129 catapults (trebuchets) hurling gunpowder bombs were decreed standard equipment on all warships, and between 1132 and 1183 a great number of treadmill-operated paddle-wheel craft, large and small, were built. These included stern-wheelers and ships with as many as 11 paddle-wheels a side (the invention of a remarkable engineer Kao Hsüan [Gao Xuan]), and in 1203 some of these were armoured with iron plates (the design of another outstanding shipwright Chhin Shih-Fu [Qin Shi-Fu]). Thus it was that the navy of the Southern Sung [Song] held off the Chin [Jin] Tartars and then the Mongols for nearly two centuries, gaining complete control of the East China Sea. Its successor, the navy of the Yuan, was to control the South China Sea also – and that of the Ming the Indian Ocean itself.

From the Yuan to the Chhing [Qing]

Under the Mongolian rule of the Yuan dynasty naval operations were particularly prominent. The campaign of 1277 involved large fleets on both sides, and in the final naval battle two years later near Canton, which had been the last temporary Sung [Song] capital, no less than 800 warships were captured by the Mongols. But all this was but the beginning of the naval activities of the unexpectedly sea-minded Mongol government. At the same time as the war against the Sung in South China, Kublai Khan's urge for world dominion was impelling him to engage in a series of formidable expeditions against Japan. In that of 1274 the fleet was composed of 900 warships, which transported a quarter of a million soldiers across the sea. In 1281 a larger armada of 4400 ships set sail, but each time the Japanese, aided by typhoons and bad weather, succeeded in repulsing the invaders and inflicting very great losses on them. The emperor was dissuaded by popular disapproval from mounting a third attack in 1283, but undeterred he despatched fruitless expeditions to Champa, Java and the Liu-Chhiu [Liu-Qiu] Islands. Unfortunately the historical sources recording these events of the Yuan period have never been investigated from the point of view of nautical technology, and much may be expected when this attempt is made.

When the Sung [Song] empire was finally conquered, the sailors in the Mongol service were called upon to perform a new task, the shipment in guarded convoy of grain supplies from the southern provinces to the northern capital, near modern Peking [Beijing]. Until the Grand Canal could be remodelled to cope with the now greatly increased amount of transport, the stability of the new dynasty depended upon the success of an alternative route. So effective did this new sea-route become that an acrimonious controversy arose between the supporters of the 'blue sea' and canal routes; this lasted for fifty years, much longer indeed than the rule of the Great Khan.

The first success of the naval service occurred in 1282. A fleet of 146 vessels was gathered by two former privateer commanders, Chu Chhing [Zhu Qing] and Chang Hsüan [Zhang Xuan] who had joined the Mongol forces in the coastal campaign against the Sung [Song]. After wintering in a Shantung

[Shandong] port they unloaded some 3280 tonnes of grain at the mouth of the river Wei near modern Tientsin [Tianjin]. Very soon the grain transported became equal to that carried by the inland waterways, some 20 100 tonnes, but party politics intensified, especially after the loss of a great grain fleet in a typhoon in 1286, and Chu and Chang were removed from their commands while work on the canal was pushed on more energetically. Nevertheless throughout the dynasty the sea route remained the more effective, and in 1291 the two old pirates, now admirals, regained control of it. Although they did not long survive the death of the Great Khan, their successors, the Muslims Qobis and Muhammad (Ho-Pi-Ssu [Ho-Bi-Si] and Ma-Ha-Mo-Tê [Ma-Ha-Mo-De]) carried the service to still greater efficiency, reaching a record annual shipment of 251 000 tonnes in 1329. Thereafter the sea transport declined, partly because of the use of canals, partly because of foreign pirates, and with the coming of the Ming dynasty the capital shifted again to Nanking [Nanjing]. Even in later times, after its return to Peking [Beijing] in 1409, the sea route never regained the prominence it had in the days of the Yuan navy.

In 1958 countryfolk planting lotus seeds some 300 km from Chinan [Jinan] came across a fourteenth-century ship in the deep mud of a tributary of the Yellow River. The vessel is typically Chinese, very long and narrow; it has 13 bulkheads and is about 20 metres in length and 3 metres wide. One can make out an emplacement for the slung rudder and the remains of two masts. The boat seems built for speed, and from the remains of helmets and other accoutrements in her, it is thought that she was a government patrol boat of the naval police on the Grand Canal and associated waters. Though not one of the greater craft of the time, this relic is of considerable interest because it is contemporary with the Catalan world-map and only a few decades later than the time when Ibn Battutah was in China.

Of what was accomplished by the peaceful maritime expeditions of the Ming in the early fifteenth century we shall have something to say in the next chapter, but a few details may well be mentioned here about the shipbuilding aspect of that remarkable navy which Chêng Ho [Zheng Ho] commanded, and which may have influenced Europe more than has generally been supposed. The shipyards, mostly on the Yangtze [Yangzi] near Nanking [Nanjing], were at the height of their activity between 1403 and 1423. Their first order was for 250 vessels, many larger than ever previously built. Authority varied, sometimes it was the Defence Production Board, sometimes the Ministry of Works. Other yards were also hard at work, and between 1405 and 1407 those in Fukien, Chekiang and Kuangtung [Fujian, Zhejiang and Guangdong] produced no less than 1365 ships of various sizes. In 1420 large-scale naval architecture attained the status of a Board of its own, the chief designer and builder being Chin Pi-Fêng [Jing Bi-Feng]. Taoists [Daoists] were appointed to select fortunate days for laying down ships, and there were offices for the various kinds of work, organising the carpenters, metal-workers, and so on. The best artisans, selected by examination, were transferred from other duties, such as the building and repair of palaces and temples. Thirteen provinces contributed by special taxation. As for the size of the junks built in these 'Treasure-ship Yards', the 62 largest were 135 metres long, and at broadest beam almost 55 metres. Each one carried a crew of 450–500 men. The poop had three superimposed decks, and there were several decks below the main one; no less than nine masts were stepped in the largest.

The size of these vessels constitutes a cardinal problem in naval archaeology. In the general anxiety to reduce the dimensions, some have suspected that the beam figure may have included the overhang of the sails when set furthest out, but this is quite unlikely. What is more to the point is the fact that in the typical Chinese build the upper decks and poop can over-ride the bottom timbers by some 30 %, so that for the dimensions given, a bottom length of about 95 metres could be assumed, and individual timbers up to 24 metres. They were certainly very broad-beamed, though none of the sources we have give the draught. In all there seem to have been 23 sizes of vessel, ranging from the largest nine-masters down to small single-masted vessels of about one-tenth the length, and whenever a figure across the beam is given, the ratio between length and breadth always remains about the same.

The sizes of the ships suggest a burden of some 2540 tonnes and a displacement of about 3150 tonnes to one scholar. Another, however, obtains much lower values – a burden of some 510 tonnes and a displacement of no more than 800 – still much greater, however, than that of contemporary Portuguese ships. This last figure, it is claimed, fits in with the numbers of the crew and marines carried on board the individual ships of Chêng Ho's [Zheng Ho's] fleets. All the same, even the lower figure cannot, it is agreed, be taken as a firm upper limit, and it is thought that perhaps some Sung [Song] ships had been larger, having 1270 tonnes burden, as mentioned in the *Mêng Liang Lu* [*Meng Liang Lu*] (Dreaming of the Capital while the Rice is Cooking [a description of Hangchow [Hangzhou] towards the end of the Sung]) of 1275. This would certainly agree with Marco Polo's evidence of almost the same time, hard though that is to interpret exactly.

A startling new development occurred in 1962 when an actual rudderpost of one of Chêng Ho's [Zheng Ho's] treasure-ships was discovered at the site of one of the Ming shipyards at Nanking [Nanjing]. This great timber is 11 metres long, of 38 centimetres diameter, and shows a rudder attachment length of 6 metres. Assuming the usual Chinese length–breadth ratio for the rudder blade, this means an area of no less than 42 square metres (Fig. 201). Using accepted formulae, the approximate length of the ship on which it has been used comes out between 146 metres and 163 metres depending on different assumptions about the draught of the vessel. This rudder-post shows

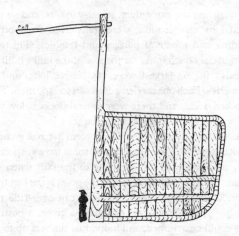

Fig. 201. A reconstruction of the rudder of one of the great ships of Chêng Ho (Zheng Ho) (*c.* 1420) based on the dimensions of an actual rudder-post recovered in 1962 near Nanking [Nanjing]. The size may be judged from the human figure.

that the Ming texts are not spinning a yarn when they give dimensions at first sight hard to believe for the flagships of Chêng Ho's fleets.

Before 1450 came a fundamental change in policy, as we shall see (p. 145). The anti-maritime party at court, for reasons still somewhat obscure, got the upper hand and the long-distance navigations were brought to an end. That it never completely destroyed the traditions of the sea, however, is indicated by the fact that in 1553 a full-dress history of Chêng Ho's [Zeng-Ho's] shipyards was written. This was the *Lung Chiang Chhuan Chhang Chih* [*Lung Jiang Chuan Chang Zhi*] (Record of the Shipbuilding Yards on the Dragon River) already mentioned, and from which Fig. 202 is taken. Written by Li Chao-Hsiang [Li Zhao-Xiang], it must be regarded as one of the treasures of Chinese technological literature. Besides a brief history of shipbuilding during the Ming dynasty, and an account of the officials who were entrusted with its organisation, the book contains descriptions and illustrations of ships built. Then not only are the yards described and plans given (Fig. 202) but there also follow specifications and dimensions of materials, tabulated with costs, as well as details of the number of shipwrights and workmen required for each particular job. The last chapter gives one of the best collections of literary and historical references to ships and shipping in all Chinese literature.

In its heyday, about 1420, the Ming navy probably outclassed that of any other Asian nation at any time in history, and would have been more than a match for that of any contemporary European State or even a combination of

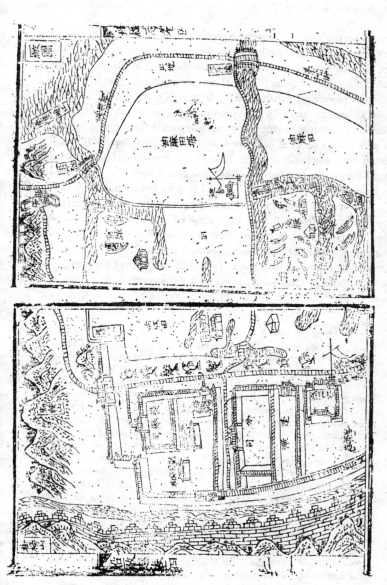

Fig. 202. One of the plans of the shipyards near Nanking [Nanjing] where the great ships of Chêng Ho's [Zheng Ho's] fleets were built and fitted out, from the *Lung Chiang Chhuan Chhang Chih* [*Long Jiang Chuan Chang Zhi*] (Records of the Shipbuilding Yards on the Dragon River) (near Nanking) of 1553.

them. Under the Yung-Lo emperor it consisted of some 3800 ships in all, 1350 patrol vessels and 1350 combat ships attached to guard stations or island bases, a main fleet of 400 large warships stationed at Hsin-chiang-khou [Xinjiangkou] near Nanking [Nanjing], and 400 grain-transport freighters. In addition there were more than 250 long-distance 'Treasure-ships' or galleons, the average complement of which rose from 450 men in 1403 to 690 in 1431 and certainly overstepped 1000 in the largest vessels. A further 3000 merchant-men were always ready as auxiliaries, and a host of small craft did duty as despatch-boats and police launches. After the great reversal of policy the navy declined very much more rapidly than it had grown, so that by the middle of the sixteenth century almost nothing was left of its former grandeur.

With the seventeenth and eighteenth centuries we reach the period of intensified intercourse with the West and the end of our archaeological tour. Indeed, in the eighteenth century Europeans began to see a good deal of Chinese ships in practical use. Sometimes also they made use of the services of Chinese shipwrights. In his voyage from Macao to explore the north-west coast of America in 1788, John Meares took a company of them along with him. Of course

Caption for fig. 202 *(cont.)*

˙The view of part of the yards is taken looking approximately south at the strip of land between the walls of the city of Nanking on the left and the Chhin-huai [Qin-huai] River debouching into the Yangtze [Yangzi] at the bottom on the right (as the legend at the top on the right says). This river, coming from Chiang-ning [Jiangning], flows all round the south of Nanking outside the city walls, sending a loop through the southern quarter; it got its name from the belief that it had first been canalised in Chinn [Qin] times, and if not salubrious it was famous for the painted boats of sing-song girls which were moored along the loop end to end. But at its mouth it was big enough to float the great hulls of the ocean-going Treasure-ships of the fifteenth century.

At the top of the plan is Ma-an Shan, a hill now within the city walls, and to its left, inside them, Kua-pang Shan [Guabang Shan] is labelled, the Hill of Hanging up the Pass-Lists of Successful Candidates. In the left-hand half of the picture from top to bottom we can make out first the Main Gate (Ta Mên [Da Men]), then the Intendant's Headquarters (Thi-Chü Ssu [Ti-Ju Si]), the Foremen's Offices (Tso Fang [Zuo Fang]), various administrative sections (Fên Ssu [Fen Si]), the Sail Loft (Phêng Chhang [Peng Chang]) and the Naval Liaison Command (Chih-Hui Chü [Zhi-Hui Ju]) marked by a flag. All around are wide fields in which hemp was grown to yield oakum for caulking. In the right-hand half of the picture two shipyards are seen with their slipways and docks, the Chhien Chhang [Qian Chang] above and the Hou Chhang [Hou Chang] below; between them there is a Guard Post (Hsün Shê [Xun She]) marked by another flag. The entrances of the channels are crossed by two floating bridges, the smaller (Hsiao Fou-Chhiao [Xiao Fou-Qiao]) above, and the larger (Ta Fou-Chhiao [Da Fou-Qiao]) below; these carry the road along the bank of the Chhin-huai River.

the Chinese artificers in this branch had not the least idea of our mode of naval architecture. The vessels of their nation, which navigate the China and adjacent seas, are of a construction peculiar to them. In vessels of a thousand tons burthen, not a particle of iron is used; their very anchors are formed from wood, and their enormous sails made of matting. Yet these floating bodies of timber are able to encounter any tempestuous weather, hold a remarkable good wind, sail well, and are worked with such facility and care, as to cause the astonishment of European sailors.

Such was the opinion of a great English navigator at the end of the century.

3

Voyages and discoveries

'Westerners' wrote the French historian Cordier in 1920, ' have singularly narrowed the history of the world in grouping the little they knew about the expansion of the human race around the peoples of Israel, Greece and Rome. Thus have they ignored all those travellers and explorers who in their ships ploughed the China Sea and the Indian Ocean, or rode across the immensities of Central Asia to the Persian Gulf. In truth the larger part of the globe, containing cultures different from those of the ancient Greeks and Romans but no less civilised, has remained unknown to those who wrote the history of their little world under the impression they were writing world history'. And the famous Edward Gibbon, as we saw (p. 2), was unaware that the Chinese had visited the seas of the southern hemisphere. Here then, within the compass of this chapter, we must do something to redress the balance.

It has been suggested that Asian sailors never rounded the Cape of Good Hope because of want of courage rather than of technical equipment. Assuming for the moment that they did not, it is extremely doubtful whether either of these propositions is true in any sense. The sewn ships of Arabia and India were doubtless too uncertain for really long voyages, yet the Indonesians accomplished the colonisation of Madagascar by sea. There was much less reason, indeed none, why the great ships of China should not have discovered the west coast of Africa, and the Australian continent too, but social or political circumstances were most certainly the inhibiting factors rather than nautical technology. From Basra to Borneo, and from Zanzibar to Kamchatka west of the Bering Sea, was not an insignificant span for the furthest ranges of the Chinese flag. And the less said about courage the better, as any modern sailor would feel if he found himself invited to undertake a voyage with the same equipment and the same facilities as the Buddhist pilgrims or the Emirs of fourteenth-century Zayton (the Arab quarter of Chhüan-chou [Quanzhou], north of Amoy).

Whoever has had the good fortune both to wander on those Fukienese [Fujianese] and Cantonese shores that saw the passage of the great ships of China, and to stand also on that hill which overlooks the Tower of Belem at the entrance to the Tagus from whence Vasco da Gama set sail, cannot but be powerfully impressed by the strange fact that the great Portuguese and Chinese voyages of discovery happened at the same time. It is indeed an extraordinary historical coincidence, that Chinese long-distance navigation from the Far East reached its high-water mark just as the tide of Portuguese exploration from the Far West was beginning its spectacular flow. These two great currents almost met, and in a single region, the coasts of the African continent. Their inspiration came from two equally extraordinary men active in maritime affairs, on the one side a royal patron of mariners, Henry the Navigator, on the other an imperial eunuch, ambassador and admiral, Chêng Ho [Zheng Ho]. The contrast is inescapable, for this was the peak of Chinese maritime enterprise.

The Admiral of the Triple Treasure

A reference to the exploits of Chêng Ho [Zheng Ho] has already been made when we discussed Chinese map-making (volume 2 of this abridgement), but it will be as well if we quote once more from the *Li-Tai Thung Chien Chi Lan* [*Li-Dai Tong Jian Ji Lan*] (Essentials of the Comprehensive Mirror of History) compiled by a group of scholars under imperial order in 1767.

In the third year of the Yung-Lo reign-period (1405), the Imperial Palace Eunuch Chêng Ho [commonly known as the 'Three-Jewel Eunuch', a native of the province of Yunnan], was sent on a mission to the (countries of the) Western Oceans.

The emperor Chhêng Tsu [Cheng Zu], under the suspicion that (his nephew) the (previous) Chien-Wên [Jian-Wen] emperor might have fled beyond the seas, commissioned Chêng Ho, Wang Ching-Hung [Wang Jing-Hong] and others, to pursue his traces. Bearing vast amounts of gold and other treasures, and with a force of more than 37 000 officers and men under their command, they built great ships [62 in number], and set sail from Liu-chia Kang [Liujiagang] ... whence they proceeded ... to Chan-Chhêng [Zhancheng] (Champa, Indo-China), and thence on voyages throughout the western seas.

Here they made know the proclamations of the Son of Heaven, and spread abroad the knowledge of his majesty and virtue. They bestowed gifts upon the kings and rulers, and those who refused submission, they overawed by the show of armed might. Every country became obedient to the imperial commands, and when Chêng Ho turned homewards, sent envoys in his train to offer tribute. The emperor was highly gladdened, and after no long time commanded Chêng Ho to go overseas once more and scatter largesse among the different States. . . .

Chêng Ho was commissioned on no less than seven diplomatic expeditions, and thrice he made prisoners of foreign chiefs. His exploits were such as no eunuch before him, from the days of old, had equalled. At the same time, the different peoples, attracted by the profit of Chinese merchandise, enlarged their mutual intercourse for purposes of trade, and there was uninterrupted going to and fro.

Thus it came to pass that in those days 'the Three-Jewel Eunuch who went down into the West' became a proverbial expression; . . . .

In this interesting summary we see at the outset some of the primary motives of the voyages. There was the search for the deposed emperor, but overshadowing it was the clear desire to impress upon foreign countries even beyond the limits of the known world the idea of China as the leading political and cultural power. There was also the encouragement of overseas trade. Had not one of the Sung [Song] emperors, Kao Tsung [Gao Zong], the founder of the Chinese navy, said 'The profits from maritime commerce are very great. If properly managed they can amount to millions (of strings of cash). Is this not better than taxing the people?' That had been about 1145, when in falling back to Hangchow [Hangzhou] the government had first become fully conscious of the importance of sea power. But it was no less applicable in the early fifteenth century when Tamerlane had just completed his general devastation of Western Asia, and all the lands and routes of Turkestan were closed again to Chinese commerce.

The long-distance voyages involved at least three specialised activities. On the naval side was the conduct of large fleets of junks, then the greatest vessels afloat, over many thousands of kilometres to regions where no organised Chinese fleets had been before. They had to be worked safely in and out of little-known ports and harbours, with a great deal of handling in the narrow waters of the South East Asian archipelago, as well as direct passages on the high seas from Malaya to Africa. On the military side there was the organisation of marines and gunners at sea and ashore, with commanders who proved efficient and successful in certain unexpected actions, though the troops were primarily ceremonial. As for the diplomatic or prestige function, what it involved in practice was giving rich presents to rulers into whose domains the Chinese ventured, at the same time inducing those very rulers to acknowledge the overlordship of the Chinese emperor, and to despatch, if possible, tribute bearing missions to the Chinese court. Under the heading of tribute a great deal of state trading was carried on, and besides there may have been some desire to foster private traders and merchants. Lastly there was what we may call a proto-scientific function. An increase in knowledge of the coasts and islands of the Chinese culture-area was required by the administration, as well as a survey of the routes to the Far West. Furthermore, there was the search for all kinds of rarities, and gems, minerals, plants, animals, drugs and the like were to be collected for the imperial museum.

One gains the impression that the more the voyages of exploration developed and the further they reached out, the more important became the collection of natural curiosities and the less important the matter of securing of tribute from local rulers, while the search for the missing emperor faded into the background. We shall come back to these aims when we compare them with those of the Portuguese pioneers.

The seven expeditions from China extended progressively westwards. The first (1405 to 1407) visited Champa, Java and Sumatra and then, going the other way, Ceylon and Calicut on the west coast of India. The second (1407 to 1409), under another commander, Chêng Ho [Zheng Ho] being absent, visited Siam and added Cochin to its Indian ports of call. On the third the fleet went to all the usual places in the East Indies using Malacca as a base, adding Quilon in south-west India, and became involved in affairs on the island of Ceylon (1409 to 1411). At this time a third able leader, the eunuch Hou Hsien [Hou Xian], joined Chêng Ho and Wang Ching-Hung [Wand Jing-Hong].

India was left behind by the fourth expedition in 1413 to 1415, which took different directions. While some squadrons visited all the East Indies again, others (based on Ceylon) explored Bengal, the Maldive Islands, and reached Iran. Such interest was roused in the Arabic culture-area at this time, including the Arab city-states on the East African coast, that in 1416 a remarkable flow of ambassadors converged on Nanking [Nanjing]. As a result, a fleet greater than ever was organised the following year to take them home again. Between then and 1419 while its Pacific squadrons went as far as Java, the Ryukyu islands and Brunei, its Indian ones ploughed the seas from the Iranian seaboard to Aden, visiting many places on the coast around Somalia and Kenya. This fifth expedition was the time when giraffes went back to Peking to intrigue the court and delight Chinese naturalists. The sixth expedition (1421 to 1422) covered the same ground but, to do so, must have split up into separate groups.

It was after this, in 1424, that there came the first warning of the calamities about to befall the Ming navy. The Yung-Lo emperor died and his successor Jen Tsung [Ren Zong], who sided with the anti-maritime party, countermanded the voyage planned for that year. But almost at once he also died, and his successor Hsüan Tsung [Xuan Zong] was left to preside over the last and perhaps the most dazzling of the Grand Treasure-Ship Fleets. It left in 1431 and before it returned in 1433 its commanders, with their 27 550 officers and men, had established relations with more than 20 realms and sultanates from Java westwards through the Nicobar Islands to Mecca in the north and the coast of East Africa in the south. Precisely how far down this coast the Chinese penetrated is uncertain – we shall return to this point shortly – but as far as their visits to the Persian Gulf and Red Sea are concerned, it should be remembered that these were not new in the fifteenth century. Chinese ships had been frequenting those waters for a thousand years. What was new was the appearance of organised naval forces with junks of great size, not simply small and isolated merchant-ships. In view of this the fundamentally peaceful character of the Ming voyages seems all the more remarkable.

Two years after the return of Chêng Ho [Zheng Ho] and his last ships the Hsüan-Tê [Xuan-De] emperor died, and the Chinese Admiralty was doomed at last. Ying Tsung [Ying Zong] and his successors listened to the Confucian 'agriculturists', the scholar-landlords, so that official maritime activities were reduced to the minimum needed to protect coast and grain-ships from Japanese pirates. This was a decision which had far-reaching results not only for the Chinese but also for world history.

It is clear that the Grand Fleets separated into a number of squadrons with particular missions. Yet their surprising activities were only the culmination of naval diplomatic missions which had been increasing since the end of the Yuan (mid fourteenth century) and which paralleled missions to western countries by land. A notable figure here was Hou Hsien [Hou Xian], who went to Tibet, Nepal and Bengal, and was considered the most important diplomat after Chêng Ho [Zheng Ho]; he was soon followed by the others.

The expeditions of Chêng Ho's time had considerable influence on Chinese literature, parallel in a smaller way to the spread of knowledge of the Portuguese discoveries through Europe. Some manuscript books of sailing directions (but no maps) have come down to us and, in addition, a number of 'portolans' or route-maps in a distinctively Chinese style appeared in the *Treatise on Armament Technology* (1628). These last derived from cartographers attached to Chêng Ho. The Great Voyages also provided material for one of the famous Ming novels, the *Hsi Yang Chi* [*Xi Yang Ji*] by Lo Mou-Têng [Lo Mou-Deng] which was published in 1597. Though it contains much fabulous material, it is nevertheless a source of reliable information on the organisation of the tribute missions and their gifts, together with interesting technical details.

China and Africa

Chinese relations with East Africa were far older than the days of Chêng Ho [Zheng Ho]. From ancient Egyptian times onwards there had been trade down the coast, and in the eighth century AD, Arab trading centres such as Mogadishiu in Somaliland and Sofala south of the Zambezi River were established. From here Arab exploration spread out, Madagascar and the

Comoro Islands in the Mozambique Channel being visited in the ninth century. What is much more unexpected is that descriptions of this part of the world can be found in Chinese literature as early as about AD 860. When Tuan Chhêng-Shih [Duan Cheng-Shi] was compiling his *Yu-Yang Tsa Tzu* [*Yu-Yang Za-Zi*] (Miscellany of the Yu-Yang Mountain Cave) at this time he included an interesting passage on Berbera, the south coast of the Gulf of Aden. The area is described with increasing detail in later books and in 1225 it was included in a text, the *Chu Fan Chih* [*Ju Fan Zhi*] (Records of Foreign Peoples), which also gives an elaborate description of the Somali coast. By 1178 Madagascar was described at some length, and the next century the whole of the coast between the Juba River in Somalia and the Mozambique Channel was described by Chao Ju-Kua [Zhao Ru-Gua]. A hundred years later, all these regions had become well known to the Chinese.

How many Chinese merchants and sailors themselves actually visited these places between the eighth and fourteenth centuries we have no means of telling. Apart from texts like those just mentioned, there is only the mute evidence of Chinese objects scattered up and down the coast. And these are many indeed, so many that it is rather hard to believe that they all came through the hands of intermediate traders. Before we briefly consider them, one positive testimony must be given and that from an Arabic source, of the presence of Chinese merchants on the shores of twelfth century East Africa. The great Sicilian Muslim geographer al-Idrisi, writing about 1154, says:

> Opposite the coasts of Zanj are the Zalej Islands [believed to be the Mafias off the Tanzanian coast about 240 km south of Zanzibar]....
> One of these isles is called ... al-Anjeb.... This island is very populous, with many villages and domestic animals; rice is grown there. There is much commerce, and markets to which all kinds of things for sale and use are brought. It is said that once when the Chinese affairs were troubled by rebellions, and when tyranny and confusion became intolerable in India, the Chinese moved their commercial centre to Zalej and the other islands which belong to it, entering into familiar relations with the inhabitants because of their equity, uprightness, amenity of customs and aptitude for business. This is why the island is so populous and so frequented by strangers.

Here we have only a glimpse, for it is not quite clear what al-Idrisi had in mind. The Chinese rebellion to which he refers sounds like that of Huang Chhao [Huang Chao] (AD 875 to 884), during which the Arab quarter of Canton was destroyed, but trouble on the East African mainland would have been a much more likely cause of the removal of Chinese trading stations to an island. Nor is al-Idrisi's reference to India easily understood. Nevertheless we may accept what he says as a picture of such activities about AD 1000. If there was one such Chinese station on the coast in Sung [Song] times, there were probably several, and merchant-junks too, to connect them with home.

Among the things which the Chinese wanted from Africa were elephant tusks, rhinoceros horns, strings of pearls, aromatic substances, incense gums and the like. Statistics preserved in the *Sung Shih* [*Song Shi*] (History of the Sung Dynasty) show that these imports increased ten-fold between 1050 and 1150. Al-Idrisi, on the other hand, tells us what Aden (and hence the Coast) received from India and China – iron, damascened sabres, musk and porcelain (typical Chinese exports), saddles, 'velvety and rich textiles' (probably silk), cotton goods, aloes, pepper and South Sea spices. Fortunately some of this was hardware and has survived until today. In 1955 one archaeologist, Mortimer Wheeler, wrote: 'I have never seen so much broken china as in the last fortnight between Dar-es-Salaam and the Kilwa Islands; literally fragments of Chinese porcelain by the shovelful. . . .'

Archaeological research in East Africa is now in full swing, and general conclusions can only be provisional. But already there are some extraordinary acquisitions. Along the entire Swahili coast to Cape Delgado 'an unexpected and improbably large quantity of Chinese porcelain' has been discovered, whole pieces sometimes being found inset in the plastered walls of houses and mosques, where also there are niches designed to contain them. A pillar tomb near Bagamoyo (opposite Zanzibar Island) was decorated with sea-green bowls of the Yuan period, exactly contemporary with the description of Wang Ta-Yuan [Wang Da-Yuan]. Nor are the finds restricted only to coastal areas, for many pieces have appeared far inland. Broadly speaking (and as might perhaps be expected) the oldest periods are represented most strongly in the north, while further south evidence points to a great upsurge of the import of Chinese goods from the middle of the fourteenth century onwards. Perhaps this may be due to the decline of kilns in the Middle East after the collapse of the Abassid caliphate in the Mongol invasions of the thirteenth century. In any case the products of Chinese culture are celebrated in Swahili literature.

The other kind of Chinese hardware on the East African coast is monetary – coins and coin-hoards. Out of a total of 506 foreign coins found on the coasts of Kenya and Tanzania and dating from before 1800, no less than 294 are Chinese, and the great majority of them, curiously enough, are of the Sung [Song] period. This may not mean that trade was more intense then than at other times, but only that African merchandise was for a period bought with money rather than bartered. The earliest coins date from about AD 610.

It is thus clear that before the appearance of European ships in the Indian Ocean, Chinese trading influence extended down the eastern coast of Africa almost as far as Natal, certainly to the mouth of the Zambesi, and that the Mozambique Channel was ploughed by Chinese hulls. How far south the

Ming fleets carried their planned investigations is, however, uncertain. Maps in the *Treatise on Armament Technology* end just south of a port called Ma-Lin-Ti [Ma-Lin-Di] but mark Mombasa above it. Since the whole coast was known to the Portuguese later on as Melinde, it is probable that Ma-Lin-Ti on this chart means not modern Malindi in Kenya but rather Mozambique (latitude 15 °S.) Again a place called Chhi-Erh-Ma [Qi-Er-Ma] may have been Kilwa (10 °S.), south of Zanzibar. Moreover, the official Chinese history names two places as being at the most extreme distance from China, saying that the Admiral (or some of his lieutenants) went there, but that no tribute missions were ever sent by the administration. These were Pi-La [Bi-La] and Sun-La. Though almost certainly on the African coast, they cannot as yet be identified, unless Sun-La be taken to mean Sofala. Since this was an Arab trading-centre it is quite likely that a squadron of Treasure-Ships went down that far, and if it did, the position reached was 20 °S.

Early Chinese knowledge of Africa is confirmed by evidence of a quite different kind. In the *Yü Ti Tsung Thu* [*Yu Di Zong Tu*] (General World Atlas) of Shih Ho-Chi [Shi Ho-Ji], printed in 1564, South Africa was shown with the correct shape, its tip pointing to the south. This is impressive because the European map-making tradition before the Portuguese discoveries was to have it pointing to the east. In fact this atlas was derived from an earlier sixteenth-century atlas, the *Kuang Yü Thu* [*Guang Yu Tu*] (Enlarged Terrestrial Atlas) which was itself based on the work of the great Chinese map-maker Chu Ssu-Pên [Zhu Si-Ben] whose work was finished a few years after 1312. The significant point here is that Chu Ssu-Pên drew it correctly too. Moreover, he was not the only Yuan geographer to do so; there was also Li Tsê-Min [Li Ze-Min] and the Buddhist monk Chhing-Chün [Qing-Jun], one working about 1325 and the other about 1370. Their work was combined in Korea to give the *Map of the Territories of the One World and the Capitals of the Countries in Successive Ages*. Here, then, in 1402, before first Portuguese caravel had sighted Cape Nun opposite the Canary Islands, Africa was made to point south with a roughly correct triangular shape, and some 35 place names including Alexandria were marked upon it. This world-map (Fig. 203) is greatly superior to the Catalan Atlas of 1375 and even to the map by Fra Mauro of 1459, presumably because the knowledge of Europe and Africa which Chinese scholars obtained from their Arab informants was better and more abundant than all that Marco Polo and other Western travellers could bring home about East Asia. The Chinese were in fact a good century ahead.

At the outset of this chapter we acquiesced in the conventional view that the Portuguese were the first to go round the Cape of Good Hope. But Fra Mauro's atlas carries two amongst its many inscriptions which in this context are very curious. In a scroll on the East African coast near Diab (the Cape), the first of these says:

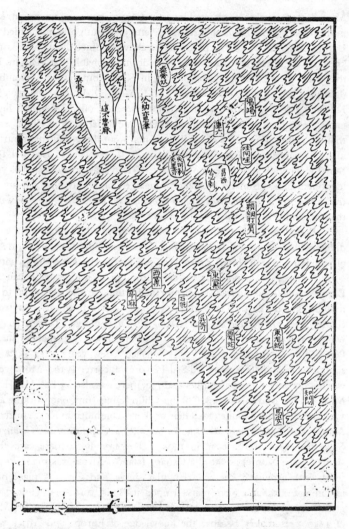

Fig. 203. The representation of the southern part of Africa in the world atlas *Kuang Yü Thu* [*Guang Yu Tu*] (Enlarged Terrestrial Atlas) by Chu Ssu-Pên [Zhu Si-Ben], of about 1315, first printed about 1555 (edition of 1779). Already in Chu's time the Chinese knew that the continent pointed due south and not eastwards. It is interesting to find that the centre of the land-mass is shown as a vast lake, probably because the existence of one or more of the great East-Central African lakes was known. Various places are named; the words Chê-pu-lu-ma [Zhe-bu-lu-ma] undoubtedly stand for the 'Mountains of the Moon' of Ptolemy; probably the modern Mt Ruwenzori on the Uganda-

About the year 1420 a ship or junk of the Indies passed directly across the Indian Ocean in the direction of the Men-and-Women Islands beyond Cape Diab, and past the Green Islands and the Dark (Sea), sailing (thereafter) west and south-west for 40 days and finding nothing but air and water. According to the estimate of her (company) she travelled 32 000 km. Then, conditions worsening, she returned in 70 days to the aforesaid Cape Diab. When the sailors went on shore to satisfy their needs, they saw an egg of the bird called roc, the which was the size of the belly of an amphora; and so great was the size of the bird that the span of its wings was 60 paces. This bird, which can carry off an elephant with ease, as well as other large beasts, does a great deal of harm to the inhabitants of those parts and is extremely rapid in flight.

Fra Mauro continues in another, more southerly, scroll, as part of a passage maintaining the continuity of the Indian and Atlantic Oceans:

Moreover I have had speech with a person worthy of belief who affirmed that he had passed in a ship of the Indies through a raging storm 40 days out of the Indian Ocean beyond the Cape of Sofala and the Green Islands more or less south-west and west. And according to the calculations of her astronomers, his guides, this person had sailed 32 000 km. Whence assuredly we may take him as sincere as those who say they have sailed 70 000 km (down the west coast of Africa and back) [i.e. the Portuguese explorers whose charts Fra Mauro says earlier in the same inscription that he had at his disposal].

This is all we know. A porthole opens to disclose some sea-going junk flying before the wind in the Agulhas Current round the Cape, then caught in the south-east trade-winds till finding no land her master comes down again into the westerlies and their current further south. And so he finds himself back again in the Indian Ocean where at some landfall his crew stumble upon the traces of giant birds. Then quickly the view is shut off. But we need have no doubt that the junks of Chêng Ho [Zheng Ho] and, indeed, of a couple of

Caption for fig. 203 (*cont.*)
Congo frontier, possibly Mt Kilimanjaro in N. Tanzania. The large island off the east coast marked Sang-ku-nu [Sand-gu-nu], lit. 'the slaves of al-Zanj', is clearly Zanzibar or Madagascar; but the similar phrase on the west of the continent (Sang-ku-pa [Sang-gu-ba]) is puzzling, unless it is perhaps the Congo (after all Songo in Portuguese). The most difficult is the name along the great north-flowing river Ha-na-i-ssu-chin (or -chih [Ha-na- yi-si-jin (or -zhi)], though it could be the Arab al-Nilal-Azraq, the Blue Nile.

centuries earlier, accomplished this, as many competent mariners agree. Curious, too, that the Venetian friar should have mentioned 1420 for one of these wanderings. Yet if Chêng Ho had nothing to do with them, may he not in some sense be called the Vasco da Gama of China? Exactly in what sense is a question we shall try to answer a little further on.

In stating his belief that the Atlantic and Indian Oceans were continuous with no land standing between them, Fra Mauro followed the general consensus of opinion of the great Arab geographers from the ninth to the eleventh centuries. The Arabs, moreover, denied the common Western opinion that the southern hemisphere would be too hot to permit human life, even if one could get there. And, of course, they knew well of a land in the west, bordering the ocean; Oporto was familiar to them and had, for a time, been a possession of theirs during their occupation of the Iberian peninsula.

The Sea-Prince of the Five Wounds

The great epic of Portuguese maritime discovery and expansion in the fifteenth century is well known. Here, then, it is only necessary to remark that during the first half of the fifteenth century while the Portuguese were creeping down the west coast of Africa, the Chinese were examining the east coast at least as far south as Mozambique; during the second half when the Portuguese found their way round into the Indian Ocean, they met no one but Arabs and Africans, since a change of policy in China had withdrawn the Treasure-Fleets for good.

The figure behind Portuguese exploration was Henry of Avis, a figure of parallel historical importance to Chêng Ho [Zheng Ho] and now better known as 'Henry the Navigator'. A prince who realised the value of ships and could talk with the rough seafarers and fishermen who handled them, he was a medieval nobleman who after spending many years discussing matters with men of learning became convinced like them of the truth of the Arab belief that south of Africa was a cape which could be rounded. He knew, too, all that the fifteenth-century Westerner could know about the Indies and the Further Indies. His hope was that his mariners would sail round the Cape and open links with the Indian and East Indian producers of silks and spices. But they could do more, for such exploration would put them in the position of turning the Arab flank, and taking the 'Saracens' in the rear. In one sense, then, the search for the southern passage was to be the new secret weapon of Christendom in its age-long battle against the 'Infidel'.

The opening rounds were fired at Ceuta, across the Straits of Morocco and a city that was a centre of Moorish naval strength. The Portuguese took and sacked it in 1415, the year of Chêng Ho's [Zheng Ho's] return from his fourth voyage. After Ceuta ships were sent out almost every year to round the ocean, for Henry 'had a wish to know the land that lay beyond the Isles of

Canary and a cape that is called Bojador, because until that time, neither in writing nor in men's memory had it been definitely known what was the nature of the land beyond that cape.' For some dozen years they were concerned with Madeira and the Azores, but in 1426 they went to the coast of the Anti-Atlas mountains and in 1434 Cape Bojador (26 °N.). Between these dates the largest of the Chinese expeditions had been prepared, had sailed, and returned. In 1444 the Portuguese reached the mouth of the Senegal River (16 °N.) and two years later the Guinea coast (12 °N.). The year 1453 was marked by two events, one colossal blow and one apparently minor affair: Byzantium fell to the Turks, while along the African coast the Portuguese sailed in their first expedition to Guinea with the primary object of trade. In 1460 Henry died, just after Sierra Leone (8 °N.) had been reached. A lull of ten years then followed, after which the Portuguese pushed on to Ashanti (5 °N.) and crossed the equator to Lopez (now in the French Congo and 2 °S). The latitude was equal to al-Jubb, where the Chinese had passed on their parallel south-eastern journeys.

After a period of struggle between Portugal and Spain (which wanted a share in the profits of West African slaves and gold), the expeditions started again on a larger scale. A peak was reached in 1488 with Bartolomeu Dias rounding the Cape of Good Hope (about 35 °S.), and the culmination point came in 1497 when Vasco da Gama left Lisbon, went to the mouth of the Zambezi, and then entered the 'Chinese' area of Malindi, just about 50 years after the Ming navy had ceased frequenting its shores. The die was now cast, the Europeans were in the Indian Ocean for good or evil – and of the latter much.

It is of considerable interest that the former Chinese presence on coasts round the Indian Ocean was soon known, but much misunderstood, in Europe. Thus the Chinese became referred to as 'white Christians' who carried 'a certain weapon like a sword on the end of a spear' – the characteristic Chinese hand weapon, the 'gisarme'. It may well be, as some have thought, that the Arabs of al-Zanj welcomed the Portuguese at first precisely because they thought they were Chinese, and only became hostile when they found that they were Christians from the West. A comment in itself this, sad and paradoxical, on that culture which bore on its banners 'on earth peace, and goodwill towards men'.

Contrasts and comparisons

We can now compare and contrast the seafarers from East and West. From the maritime point of view – the central theme of this chapter – we can see at once that the Chinese achievement of the fifteenth century involved no revolutionary technological break with the past, while that of the Portuguese was more original. The Chinese had possessed their fore-and-aft lug-sails since

Fig. 204. Drawing of a five-masted Chinese Pei-chi-li [Bei-ji-li] freighter to give some indication of the probable type of build of the much larger Treasure-Ships of the Grand Fleet of the fifteenth century.

the third century AD at least, and already in the time of Marco Polo their ships were many-masted. If they used the mariner's compass in the Mozambique Channel, they were only doing what their predecessors had done in the Straits of Taiwan right back to the foundation of the Sung [Song] navy at the beginning of the twelfth century. Though their stern-post rudders were attached in weaker fashion to the hull than those of the Westerners, they were highly efficient in more ways than one, and descended from patterns as early as the first century AD. The most obvious difference which would have struck everyone if the vessels of da Gama had met those of Chêng Ho [Zheng Ho] lay in the much greater size of the Treasure-Ships of the Grand Fleet (Fig. 204), for many of these were some 1500 tonnes if not considerably more, while none of da Gama's were over about 300 tonnes and some were much less. In shipbuilding China was far ahead of Europe.

But while the Chinese vessels were the culmination of a long evolutionary development, those of the Portuguese were relatively new in type. At the end of the fourteenth century European ships were equipped only with square-sail rig; the *barca* might be some 30 tonnes with one mast, or up to 100 tonnes if with two. The early vessels sent out by Prince Henry were doubtless of this kind. But finding the north-east trade winds quite contrary to their return from the Guinea coast, the Portuguese threw overboard the square-sail rig, and for their famous caravels adopted the lateen or triangular sail from their enemies, the Arabs. This permitted sailing much closer into the eye of the wind. By 1436, the end of the Chinese period, these ships carried lateen sails on as many as three masts and averaged between 50 and 100 tonnes. Then, as the century went on, the superior advantages of the square-sail for running before the wind re-asserted themselves, and ships began to be built combining both rigs. Such was the *caravela redonda* of 1500, and the *Santa Maria* of Columbus. But the originality of the Portuguese seems somewhat qualified when we remember that of the basic inventions they used, the mariner's compass and the stern-post rudder were transmissions from much earlier Chinese practice, the principle of multiple masts was characteristically Asian, and the lateen sail, as we have said, was taken directly from the Arabs. Of the comments of the first Chinese sailors who examined European vessels little or nothing has remained, but from the middle of the sixteenth century there was a slight admixture of types, though Chinese shipbuilding remained for the most part unchanged.

There was another matter in which the Portuguese showed seemingly more originality than the Chinese, and this was the use of the régime of winds and currents. The problems they had to face were more difficult, and they rose gallantly to the challenge. Almost as far south as Madagascar the Chinese were in the realms of the monsoons, the 'junk-driving winds' with which they had been familiar in their own home waters for more than a millennium. But the inhospitable Altantic had never encouraged sailors in the same way, and though there had been a number of attempts to sail westwards, that ocean had never been systematically explored. First the Guinea coast proved to be a trap, for while winds and currents helped the ships down, the return journey meant endless tacking from side to side. But the Portuguese knew that above the 'Horse Latitudes' – a belt of calms and light airs bordering on the northern edge of the north-east trade-winds – there were strong westerly winds which would blow them home. Thus to return from their fortress-factory at El-Mina on the Volta, they bore out far westwards into the Atlantic and then came north to run for the Tagus. This 'Sargasso Arc' was a state secret, all the more important because the Portuguese had acquired particular skill in building and handling caravels. Indeed, their sale abroad was forbidden, and other countries constructed and sailed them only with difficulty.

The desire for secrecy meant that the records of Portuguese exploits were all kept at Lisbon, and when a large part of the city was destroyed in an earthquake in 1755 they were destroyed. Thus documentary proof for many alleged Portuguese discoveries has been lost. For example we now have no sure proof that in 1485, three years before the voyage of Bartolomeu Dias, that they did reach India. Nor is it possible to confirm the report by the Arab navigator Ibn Majid that a Portuguese expedition had been wrecked near Sofala as early as 1495; this is a pity because it raises doubt about the customary belief that da Gama was the first European to sail up the coast to Sofala.

As they ventured down the African coast, the Portuguese encountered an opposite situation. It was progress southward that was now hindered by currents and by winds, but once beyond a latitude of 35 °S., they were helped on their way. Thus another great arc was devised. Such then was the navigation of da Gama and Pedro Cabal (who is credited with the discovery of Brazil). The fact that they sailed with barques rather than caravels is highly significant, for it can only mean that previous Portuguese explorers had plotted out the route.

Now we must briefly consider war and trade. Here the contrast between the Chinese and the Portuguese is an extraordinary one, for while the entire Chinese operations were those of a navy paying friendly visits to foreign ports, the Portuguese east of Suez engaged themselves in total war. Already in 1444 the first casualties of their campaign had occurred in Mauretania, but so long as they were working down the West African coast their aggressive activities were (apart from slaving) relatively restrained. Only after 1500 when they were in a position to carry on terrorist warfare against East African Arabs, and then against the Indians and other Asians, did European naval power show what it could do in earnest. Before the coming of the Portuguese the Arab city-states had no defence works; these only arose when it became evident that it was the settled policy of the Westerners to destroy the Arab African–Indian trade root and branch. Unhappily cruelties were deliberately adopted as a terrorising policy by da Gama and other commanders. The limbs of executed Muslims were fired over Indian cities, helpless fishermen were tortured, and of those from the African coast that fell into the hands of Afonso de Albuquerque, the noses of the women were cut off and the hands of the men. Such facts as these are mentioned with reluctance, but they certainly do something to correct the stereotyped image (still met with in Europe) that Asians have been more cruel and barbarous than Europeans.

What could the Chinese show in comparison with all this? They bestowed gifts and those who refused submission they overawed by a show of armed might. On all their expeditions there were only three occasions when fighting actually took place. The first was in 1406 when a tribal chief of Palembang who had been pillaging merchants, made a surprise attack on the Chinese camp; he was defeated, captured and subsequently executed in Nanking [Nanjing]. The third happened some seven or eight years later, when one of the pretenders to the throne of north-western Sumatra quarrelled over the distribution of the Chinese gifts. He, too, led forces against those of Chêng Ho [Zheng Ho], and was captured with his family, finally being taken also to Nanking where he was executed. The second occasion was much the most serious. In 1410 the King of Ceylon enticed Chêng Ho's expeditionary guard into the interior and then demanded excessive presents of gold and silk, meanwhile sending troops to burn and sink his ships. But Chêng Ho pushed on to the capital, took the king and his weakly defended court by surprise, and then fought his way back to the coast with his captives, routing the Sinhalese army on the journey. The prisoners were taken to Nanking, where they were kindly treated and sent home again after an arrangement had been made to choose a relative of the king as his successor. Naval armed might thus meant something very different in the Chinese and Portuguese interpretations.

As for trade, our knowledge of the inner workings of that matter is as usual still very deficient, but it is clear that what was done both by the Chinese and the Portuguese was done under the aegis of their respective economic systems, and these were very different. It seems evident that the Portuguese activities were from the start much more concerned with private enterprise. The search for the 'El Dorado' which would make one's personal fortune was an inherent part of the *conquistador* mentality. The further the expeditions went, the more necessary was it to make them at least financially self-supporting. Trading ventures were certainly encouraged and licensed by the Portuguese court, but behind it stood international finance and, indeed, the developing capitalism of all Europe.

By contrast the Chinese expeditions were the well-disciplined naval operations of an enormous feudal-bureaucratic state, the like of which was not known in Europe. Their impetus was primarily governmental, their trade (though large) was incidental, and the 'irregular' merchant-mariners whose trafficking was to be encouraged were mostly humble men of small means. The bureaucracy in China generally saw to that. Only their numbers made them important. And what was true of trade in general was true also of the slave-trade in particular. The Chinese and other Asian nations had been using negro slaves for many centuries, but the fact that their slavery was basically domestic kept the practice within bounds. Not so the use of Africans in agricultural plantation labour, particularly in the New World, which brought it about that in the 160 years between 1486 and 1641, no less than 1 389 000 slaves were taken by the Portuguese from Angola alone. Their expeditions down the West Coast of Africa involved kidnapping and slave-raiding forays from the very first. This had been the mutual custom of Moors and Christians all round the Mediterranean throughout the Middle Ages, but now sadly it was extended to those who never had any part in that quarrel.

Thus the paradox emerges that while the feudal state of Portugal, hardly emerged from the Middle Ages, founded an empire of mercantile capital, bureaucratic feudalism gave to China the characteristics of an empire without imperialism. But if bureaucratic feudalism was certainly not the economy of the future, we must beware of doing injustice to the first Portuguese merchants battling in the Indian Ocean; perhaps they were caught in the mesh of economic necessity. During da Gama's first visit to Calicut in 1498 a highly significant event occurred. When the Portuguese presented the gifts they had brought – wash-basins, sugar, oil and honey, cloth and such like – the king laughed at them and advised the admiral rather to offer gold. At the same time, the Muslim merchants already on the spot, told the Indians that the Portuguese were essentially pirates, possessed of nothing the Indians could ever want, and prepared to take what the Indians had by force if they could not get it otherwise.

There is something very familiar about this scene. It symbolised a fundamental lack of balance in trade which had been characteristic of relations between Europe and East Asia from the beginning, and which was destined to continue until the industrial age of the late nineteenth century. Broadly speaking, Europeans always wanted Asian products far more than the Easterners wanted Western ones, and the only way of paying for them was in precious metals. This occurred at many places on the east–west trade-routes, but primarily of course in medieval times at the Levantine borders between Christendom and Islam. The Chinese on the other hand probably never had to face an adverse balance of trade, for everyone thought highly of silk and lacquer; it was fair exchange for anything the Chinese wanted to buy. Indeed, the Mediterranean region acted for two millennia as a kind of monstrous pump continually piping off towards the East all the gold and silver that entered it. Alexandria might pay partly with glass, medieval Western Europe partly with slaves, Venice with mirrors and England with tin, but the Arabs, the Chinese and the Indians took little interest at any time in the most typical European products such as woollen cloth or wine, and when all the barter was over, there remained a large perennial deficit.

What then were the Portuguese sea-captains to do at the end of the fifteenth century? Crusading apart, they needed spices. The huge European demand for pepper was sure and certain, and the expression of a very real need, not a luxury trade. Until the full development of winter fodder a couple of centuries later, only those animals needed for breeding or for labour could be kept during the winter-time. The others had to be killed and their meat preserved by salting, and this process needed pepper by the shipload, the pepper which now just lay within the grasp of the Portuguese – if they could pay for it. Hence the importance of gold from Guinea, which they obtained by bartering horses, wheat, wine and cheese, copper ware and other metals, blankets and strong cloth. They do not seem to have defrauded the Africans who, unlike Asians, were pleased with this exchange. Unfortunately, though, the African gold was never sufficient for their needs, so the Portuguese needed other trading bases and succumbed to the temptation by taking cities such as Malacca by sword-point. And the real criticism of their operations is that they were not content with a reasonable share of Asian trade; what they wanted before long was the complete domination of the trade and of the traders.

The last difference between the sailors of East and West was religion. Missionary activity, well-intended, accompanied Portuguese exploration from an early time. Yet by the end of the fifteenth century the war against all Muslims was being extended to all Hindus and Buddhists too, save those with whom it suited the Portuguese to arrange a temporary alliance. Then, in 1560 the Holy Inquisition was established at Goa, where it soon acquired a reputation more unsavoury than that which it had in Europe. Non-Christians as well as Christians were subjected to all those forms of secret-police terror which have disfigured our own century, yet more abominable here perhaps because enlisted in the interests of high religion.

What a contrast on board Chinese ships. Without forsaking the basic teaching of Confucianism and Taoism [Daoism], Chêng Ho [Zheng Ho] – a Muslim by birth – and his commanders were 'all things to all men'; in Arabia they conversed in the tongue of the Prophet and recalled the mosques of Yunnan, in India they presented offerings in Hindu temples, and venerated the traces of Buddha in Ceylon. Indeed, in 1411, at Galle in Ceylon they erected a votive memorial in three languages. There was a text in Chinese to the Buddhist–Taoist goddess of the sea giving thanks for preservation, one in Tamil saying that the Chinese emperor having heard of the fame of a certain incarnation of Vishnu, had caused the stone to be set up in his praise, and a third, in Persian, saying that it was for the glory of Allah and some of his (Muslim) saints. But if they are different in their dedications, the texts all describe almost identical sets of presents which, one must conclude, were brought out by sea and handed over to the representatives of the three most important religions practised on the island. Such humanistic catholicity contrasts indeed with the *autos-da-fé* of Goa later on.

The captains and the kings depart
How did it all end? Greater robbers came to prey upon the lesser. During her temporary union with Spain, the Portuguese empire remained intact, but when she shook off this unwanted association in 1640 it was too late to save her Asian dominion, which consisted essentially of a long trade route dotted with fortresses. The great eastern centres fell one by one to the Dutch, who had no high-flown ideas about the conversion of Asia to Christianity; what they minded was business. But others could play at that game, and the

Netherlands empire in turn passed to the French and then the British. With the decline of colonialism in our own time the wheel has come the full circle, and a resurgent Asia takes its rightful place in the councils of the world.

The decline of Chinese long-distance shipping had set in even faster. In China the critics had always been far more numerous and determined. The Confucian bureaucracy, with its country landlord basis, was always liable to look askance at any intercourse with foreign countries. These countries were of no interest in themselves and could offer nothing but unnecessary luxuries. But according to the classical Confucian pattern of scholarly austerity, to which national sentiment bound the court itself, at least in theory, unnecessary luxuries were deeply wrong. And since all real needs of food and clothing, including even the magnificent products of Chinese craftsmanship, were available in abundance at home, what good could it possibly do to spend money on seeking strange jewels or other things with dubious properties abroad? The Grand Fleet of Treasure-Ships swallowed up funds which, in the view of all right-thinking bureaucrats, would be much better spent on water-conservancy projects for the farmer's needs, or in agrarian financing, or the like. Indeed the Confucians were not in favour of too much aggrandisement of the Court, for in practice it meant the aggrandisement of its Grand Eunuchs. It was thus no accident that the admirals of the fleet were mostly eunuchs; in fact the whole episode of the great Chinese navigations was only one engagement in that administrative battle between Confucian bureaucrats and Imperial eunuchs which had been going on at least since Han times, and was still to continue for many years. Though modern sympathies generally lie on the side of the bureaucracy, it must be recognised that in this case at least, the eunuchs were the architects of an outstanding period of greatness in China's history.

It seems clear that Chêng Ho [Zheng Ho] and his associates must certainly have presented the fullest records of their voyages to their imperial master. But before the century ended these were destroyed by administative thugs in the service of the Confucian anti-maritime party. In his *Kho Tso Chui Yü* [*Ke Zuo Zhui Yu*] (Memorabilia of Nanking [Nanjing]), compiled in 1628, Ku Chhi-Yuan [Gu Qi-Yuan] tells us that sometime between 1465 and 1487 an order was given to search the state archives for these very documents, and when they were found the Vice-President of the War Office burnt them, considering their contents to be 'deceitful exaggerations of bizarre things far removed from the testimony of people's eyes and ears'. Other sources give further details, from which it seems that this must all have happened about 1477, just at the time that the Confucian bureaucrats were strongly opposing the plans of the eunuch Wang Chih [Wang Zhi] to restore Chinese power in South East Asia.

All this does not of course exhaust the causes of the decline of the Ming navy. Economic factors were at least as important. Very great profits accrued to China from the tribute-trade system of Chêng Ho's [Zheng Ho's] time, but by the middle of the century a severe currency depreciation had set in, the value of the paper notes falling to 0.1 % of their face value. Had the long-distance voyages continued, China would have been forced to export precious metals. At the same time there was an increase in private trade beyond what had been contemplated at the beginning of the century, and in addition there was a technological revolution. For centuries canal carriage and sea transport had competed in the essential function of grain shipment from north to south, but in 1411 the engineer Sung Li [Song Li] perfected the water supply of the summit of the Grand Canal, thus converting it into a full capacity all-seasons proposition at last. In 1415 the maritime grain-transport service was abolished.

Military events also intervened. The serious deterioration on the north-western frontiers diverted all attention from the sea, and in 1499 at the disastrous Battle of Thu-Mu [Tumu], that Chinese emperor who had suppressed the Treasure-Ship fleets was himself captured by the Mongol and Tartar armies. At the same time there was a significant shift in population from the south-eastern seaboard provinces, reversing the trend which had been so strong at the beginning of the Southern Sung [Song]. Finally, the develop-ment of a sterile conventionalised version of the neo-Confucian outlook (see this abridgement, volume 1, p. 227f), markedly idealist in metaphysics and Buddhist in religion, led to a loss of interest in geographical science and maritime techniques, replacing the energetic valour of the early Ming by an introspective culture and political lethargy. This was, indeed, one aspect of a general decline which reflected itself severely in many branches of science and technology.

The upshot of all these factors was that a great nursery of deep-water sailors was lost, and orders no longer flowed into the maritime shipbuilding yards, which were run down to maintenance level. The navy simply fell to pieces. By 1474 only 140 warships of the main fleet of 400 were left. By 1503 the Têngchow [Dengzhou] squadron had dropped from 100 vessels to 10. Desertions occurred wholesale and the corps of shipwrights disintegrated. In the sixteenth century the anti-maritime party grew ever more powerful. The shipyards contracted even further and the workers diverted to other employ-ments, while maritime activity was discouraged as much as possible. Indeed, by 1500 regulations were in force that made it a capital offence to build a sea-going junk with more than two masts. Then an edict of 1525 authorised coastal officials to destroy all ships of this kind, and to arrest mariners who continued to sail in them. Another of 1551 declared that whoever went to sea in multiple-masted ships, even to trade, was committing a crime equivalent to commu-nicating with foreigners. This was the agoraphobic mentality that closed off Japan for two centuries, and though it never prevailed in China to the same extent, the great naval possibilities had been done to death.

Before the sixteenth century was out, the policy of the short-sighted landsmen of the Ming court laid the coast open to ferocious attacks by Japanese pirates which were overcome only with the greatest difficulty. There followed an increase in naval strength which was to prove valuable at the end of that century in working with the Koreans to repel Japanese fleets. In the seventeenth century the last remains of the Ming navy fought against the Manchus and their allies the Dutch, whom the Chinese had expelled from Formosa in 1661. But the Chhing [Qing] emperors were not at all interested in the sea, and under them the navy languished. However, Chinese shipping gradually recovered and most of the traditional types of build thus came to be preserved.

Motives, medicines and masteries

It is reasonable to question whether, if the debacle of the Chinese Admiralty had not taken place, their expeditions might have continued. Would they have gone round the Cape of Good Hope? Perhaps. But the Chinese motive was never primarily geographical exploration: what they sought rather was cultural contacts with foreign peoples, even if quite uncivilised. The Chinese voyages were essentially an urbane but systematic tour of inspection of the known world. The Portuguese motives were not primarily geographical exploration either, but religious and economic as we have already seen.

On the Chinese side, it is hard to say whether the search for the dethroned emperor was ever more than a pretext; the main motive was surely the demonstration of China's prestige by obtaining the nominal allegiance (and rich exchange) of far-away princes, and this was precisely what the Confucian bureaucrats thought so unnecessary. But when one compares the Chinese and Portuguese ventures over the whole range of our knowledge of them, it does seem as if the proto-scientific function of collecting natural rarities, strange gems and animals, was more marked in the former. Before long, of course, the curiosity of Renaissance men in all exotic things asserted itself strongly in the West, but this was an old Chinese tradition too which had been very powerful, for example, in the Thang [Tang] period. Said Huang Shêng-Tsêng [Huang Sheng-Zeng]:

> Amid the thundering billows and surges reaching mountain-high, helped by their flying masts and labouring oars ... the envoys journeyed many myriads of *li*, and in their voyaging to and fro spent well-nigh thirty years. ...
> Then, their vessels filled with pearls and precious stones, with eagle-wood and ambergris, with marvellous beasts and birds –

unicorns and lions, halcyons and peacocks – with rarities like camphor and gums and essences distilled from roses, together with ornaments such as coral and divers kinds of gems, the envoys returned.

Some of the specimens had a particular value for the Chinese court; the giraffe, for example, was identified with the mythical animal *chhi-lin* [*qi-lin*] which according to age-old legend was one of the greatest auspicious natural signs of an imperial ruler of perfect virtue. Collecting went on everywhere, as the naval secretary Ma Huan recorded in a passage about Zafar in Arabia:

> When the Chinese ships arrived, after the reading of the Imperial Rescript and the offering of presents, the King despatched his chiefs all over the country to command the people to bring incense, dragon's blood (resin), aloes, myrrh, benzoin, liquid storax, *Momordica* seeds, etc. in exchange for hempen cloth, silks, and porcelain ware.

What claims our attention here is the mention of *Momordica*, because the seeds of this ornamental vine of the gourd family are used as a drug by Chinese physicians and pharmacists. Could Chêng Ho's [Zheng Ho's] men have been looking for medicinal substances? There were many able physicians among them and it may be that the search for new drugs was a more important motive than has usually been thought both for the Portuguese and for the Chinese ventures. Both in Europe and China the fourteenth century had been a time of very serious epidemics. The notorious Black Death had ravaged Europe, and similar outbreaks including bubonic and pneumonic plague had occurred in China eleven times between 1300 and 1400. It was only reasonable to think that for new and dreadful diseases new and powerful drugs must be found. That the intrepid exploration of previously unknown lands did in fact produce these is a matter of common knowledge – the coca leaves from which the Andean Indians chewed their cocaine, and above all the 'fever bark tree', or 'Jesuit's bark', from trees of the *Cinchona* genus, which was a not ineffective remedy for malaria, probably the greatest scourge of man in historical times.

In the West this new knowledge was enshrined in two important books, *Colloquies on the Simples and Drugs of India* by Garcia da Orta which was printed in Goa in 1565, and Nicholas Monardes' *Joyfull Newes out of the New-Found Worlde* (i.e. America), first printed at Seville (though with a less exciting title) in 1565. There were Chinese parallels as we might imagine. Something appeared in the last great edition of the *Chêng Lei Pên Tshao* [*Zheng Lei Ben Cao*] (Classified Pharmaceutical Natural History) which came out in 1468 (Fig. 205). There was also great interest in Arabic drugs and medical treatment as is shown by the publication in the early decades of the fifteenth century of the *Hui-Hui Yao Fang* (Pharmaceutical Prescriptions of the Muslims),

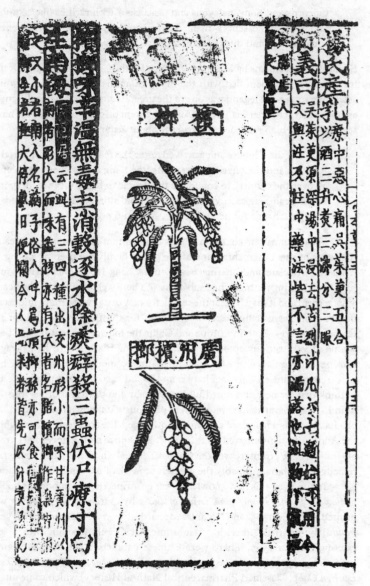

Fig. 205. To illustrate the role of natural history in the Chinese discoveries and navigations of the early fifteenth century – a page from the rare 1468 edition of the *Classified Pharmaceutical Natural History*. Here is illustrated the betel-nut tree, *Areca catechu*. Above is the whole palm, and below a drawing of the panicle or cluster with its hundred egg-sized white fruits, each

probably by an Arabic or Persian physician of the Yuan period and translated into Chinese about 1360. There is a field for further research here; some of the inorganic products discovered on these voyages can hardly have been forgotten, and in view of the very rapid spread of plants such as tobacco and maize in China after the discovery of the New World, it would be surprising indeed if the great voyages had brought no new drugs home.

The time has now come to draw all these threads together. Sofala, by the sea-routes, was just about half way between Lisbon and Nanking [Nanjing]. Might it not perhaps have changed the course of history if the first Portuguese vessels coming past Sofala to Malindi had met much greater fleets of bigger ships with company more numerous than their own – and people, too, with very different ideas about the proper relations between civilised men and barbarians? 'Coming into contact with barbarians', wrote Chang Hsieh [Zhang Xie] in 1618, 'you have nothing more to fear than touching the left horn of a snail. The only things one should really be anxious about are the means of mastery of the waves of the sea – and, worst of all dangers, the minds of those avid for profit and greedy of gain.' And indeed it was in accordance with this enlightened conception of inter-cultural contact that the Chinese set up no factories, demanded no forts, made no slave-raids, accomplished no conquests. Their total lack of any wish to make religious converts prevented any friction from that source. The governmental character helped to restrain individual avarice and the crimes to which it could give rise.

On the other hand it is clear enough that the Portuguese behaviour was a development of the Crusader mentality. They were at war. But if the naval struggle against the Muslim mercantile states on the shores of the Indian Ocean was a continuation of the 'Holy War' for the holy places, it turned insensibly into something quite different, an insatiable thirst for gold, and not only Muslim gold. There was also an obsessive desire for power over all African and Asian peoples, whether or not they had anything to do with Islam. And now too there began to come into play that decisive superiority of armaments which the scientific Renaissance was to give to Europeans, enabling them to dominate the Old World and the New for three centuries. 'From the Cape of Good Hope onwards', wrote Capt. João Ribeiro in 1685, 'we were unwilling to leave anything outside our control; we were anxious to lay our

Caption for fig. 205 (*cont.*)
containing nut and pulp within its leathery rind. The description, beginning in the first column to the left of the drawings, states as usual the basic characteristics of the product, that it is bitter, harsh and warming, that it contains no powerful active principle, that it cures or helps to cure such and such diseases, and that it grows in the lands of the South Seas. Two quotations from earlier writers follow.

hands on everything in that huge stretch of over 5000 leagues from Sofala to Japan.... There was not a corner which we did not occupy or desire to have subject to ourselves. And if China had been any weaker than she was, this vaulting ambition would not have spared her. Indeed, even the earliest Portuguese accounts of China contained estimates of her 'conquerability'.

As we have seen the 'Portuguese century' was also the 'Chinese century'. Our feeling of ambivalence towards Portugese explorers and *conquistadores* cannot be overcome. Their great and courageous actions compel all admiration. Their behaviour and policy towards the Arabs and Asians is often excused by the roughness and violence of the time. But the Chinese crews and captains were exact contemporaries of the Portuguese empire builders, and they were not warlike. So let us cherish the memories of those from the ancient area of Lusitania who were truly great, not so much Albuquerque and da Gama, but the navigators and map-makers, the astronomers and naturalists. The stature of Prince Henry is nothing dimmed; he remains for ever an inspiring and lovable figure. And from lesser men we have plenty to choose: João Fernandes who lived friendly with the Arabs and negroes of Mauritania, Fernão Queiroz who wept for the doings of his countrymen in Ceylon, to name but two. We may also celebrate Fernão Mendes Pinto who in 1614 wrote the first autobiographical novel and gave a dramatic judgement on the Deeds of his nation. Indeed, a veiled criticism of the Western attitude to Asians runs all through it, with a conviction that imperialism rested on impiety.

CHINA AND AUSTRALIA

Who first navigated the waters of Australia? We know that New South Wales was named by Captain Cook in 1770 and that Captain Dampier had explored the north-west and western coasts of the continent from 1684 to 1690. Before that a long series of Dutch surveys between 1606 and 1665, is equally established. Sixteenth century visits are less certain, though it seems quite probable that either Cristóvão de Mendonça in 1522 or Gomes de Sequeira in 1525 trod upon Australian land and met its aboriginal people. A French claim for 1503 is more shadowy. In recent times, however, the question of a possible pre-European discovery of the great island continent by Chinese sailors has been raised in serious form.

The subject is interesting partly because of the wide area of the Southern Seas over which Chinese discovery and traffic did certainly extend. For the Chinese had maritime and commercial relations with the Philippines, Java, Bali, Borneo and Sarawak, and the Moluccas and Timor, not only in the time of the great Ming expeditions, but also at least as far back as the Sung [Song]. There was traffic also with Nan Hai, the South Sea Islands or East Indies during the fourteenth century. Chinese influence in these far-flung Indonesian island countries is shown today, as in East Africa, by the presence of ceramic remains, many of high quality and great beauty. Moreover, the abundance of fine Thang [Tang] ware in various parts of Borneo demonstrates that the trade was already old in the fourteenth century, while Sarawak has dated pieces from the time of Chêng Ho [Zheng Ho] a century later.

Since Timor is only just over 640 km from Port Darwin, there seems no inherent improbability in a visit of Chinese ships to that part of the Australian coast at any time from the seventh century AD onwards. Hence the interest in a study made in recent times by the sinologist Charles Fitzgerald, who after disposing of several baseless claims, drew attention to the find of an undisputably authentic Chinese Taoist [Daoist] statuette, about 10 centimetres high, near the shore at Port Darwin. The original find was made in 1879, and the statuette was well over a metre below ground among the roots of a Banyan tree at least 200 years old which had to be removed during road making. Black with age when unearthed, the statuette is Ming or early Chhin [Qin] in style, quite reasonably contemporary with Chêng Ho [Zheng Ho]. It could, therefore, have been left before the earliest European discoveries of Australia.

That the image is Chinese is certain, but it would be hard to prove that it was left there by the crew of a Chinese junk rather than by Malay or Sundra fishermen who, like all South East Asians, have treasured cult-objects of Chinese origin. The Macassarese and Buginese used to make annual visits to the Australian coast, and written records of these are plentiful from the eighteenth century onwards. These trading visits were broken off by the Australian government in 1907, but the aboriginal inhabitants still look back at their contact with the Malays as if to a Golden Age. That the Chinese themselves were not far out of the picture, however, is shown by the fact that of all the things which the Northerners came for, the sea-slug or trepang was perhaps the most important. It is made into glutinous soups, and is a delicacy which has always been exclusively Chinese. Moreover, the Chinese alone have been successful in the preparing trade, which involves drying and smoking under mangrove-wood fires. Thus there would seem to be much significance behind the report that according to aboriginal tradition the Macassarese were preceded by a people they call the Baijini, much lighter in colour and possessing an advanced technology. If these were indeed from China, then perhaps the statuette at Darwin is a true record of their visits, and the latter half of the fifteenth century remains a possible time.

CHINA AND PRE-COLUMBIAN AMERICA

The discovery of the American continent by Buddhist monks from China in the fifth century AD was suggested by Joseph de Guignes, an eighteenth-century scholar. His claim is now discounted, but the texts from which he drew his conclusions are perfectly sound as far as they go. The basic description is in the *Liang Shu* (History of the Liang Dynasty) which dates from about AD 629; this says that in the year 499 a monk named Hui-Shen appeared in the capital and gave a circumstantial account of what he had seen in Fu-Sang, a country east of China and lying 20 000 *li* east of Ta-Han [Da Han] (the Buriat region of Siberia). He described the curious trees from which the country took its name, saying that they afforded food and that bark-cloth and a kind of writing paper were made from them. The people lived in unfortified wooden houses and were not warlike; they had oxen and horses, and drank the milk of deer. Gold and silver were not held in esteem among them, and they had copper but no iron. Hui-Shen also described their mating and funeral customs, the absence of taxes, and the cyclical changes of the colour of the ruler's vestments. He added that formerly the Fu-Sang people did not know the law of Buddha, but that in 458 five Kashmiri monks had gone there, and that since that time their way of life had much improved. He also added to his story an account of a country of Amazons, even more fabulous, which lay beyond Fu-Sang yet further to the east. The texts conclude with a statement that in 507 a Fukienese ship was blown by a storm far east into the Pacific to an island where the men had faces like dogs and the people lived mainly on a diet of small beans.

Now Fu-Sang had a long background in Chinese literature which de Guignes did not know. Texts from the Han relate various fables about it. But this does not mean that Fu-Sang was not thought of in later times as a real place. Yet no one knew where it was except that it was thought to lie somewhere far east on the shores of the Pacific. In the late nineteenth century it seemed most likely to have been the long island of Sakhalin, north-east of Japan, though Kamchatka and the Kuriles have also been considered as possibilities.

However, in 1875 some scholars went to great lengths to show that some features of Amerindian civilisations corresponded to Hui-Shen's account, though their arguments remained unconvincing. Nevertheless, there still remained a residuum of belief that connections of some kind between the Asian and Amerindian cultures had existed, though after further criticism even this was laid to rest by the time of the First World War.

In November 1947 Joseph Needham found himself in Mexico City as part of an United Nations specialist organisation, and for a couple of months had the opportunity of experiencing a culture that was not only Spanish but also profoundly Amerindian. During his stay he became deeply impressed with obvious similarities between many features of the high Central American civilisations and those of East and South East Asia. First, it seemed striking that the Central Amerindian civilisations all arose on the western side of the continent (Fig. 206). Then there was the predominance of horizontal lines in

Fig. 206. Distribution of the high cultures of Central America (after W. Krickeberg).

the terraces and monumental stairways of Central Amerindian temple and town patterns. In addition, sky-dragon motifs were to be seen everywhere, as well as those with double-headed serpents (symbolic in China of the rainbow), and there were many other similar objects from drums to terracotta figures and groups, and even paintings very similar to those of the Chhu [Chu] and Han. He also noticed similarities between the permutations used in the Maya, Aztec and Chinese calendars, in the ideographic script, as well as far-reaching parallels in the 'symbolic correlations' (colours, animals, compass-points, etc.) and even legends about the cosmos. All combined to give him an overwhelming impression of cultural influences exerted on the Amerindian cultures by those of Asia. He also found it strange that jade should have been so treasured by the Aztecs and Mayas, as dearly as by the Chinese, and even stranger that on both sides of the Pacific, jade beads or cicadas should have been placed in the mouth of the dead. His astonishment turned to conviction when he discovered that in all these civilisations the jade corpse-amulets were sometimes painted with the life-giving colour of red cinnabar or haematite.

Dr Needham also knew that in the spheres of games, divination, art forms, computing devices, etc. ethnologists had long recognised similarities with Asia. He was also aware, however, that those who had studied the

Amerindians had for many years denied any external influences on the development of such cultures. But such an accepted opinion began to change in the late 1960s, and a mountain of evidence is now accumulating that between the seventh century BC and the sixteenth century AD, that is throughout the pre-Columbian ages, occasional visits of Asian peoples to the Americas took place. We have, therefore, to recognise the arrival from time to time of small groups of men (and women also) with a background of high culture, though never any massive invasion like that of the Europeans in the late sixteenth century.

When scholars rejected the Fu-Sang story they were reckoning without the sailing raft. Ignorance of practical techniques and their development proved once again the Achilles heel of literary history. Moreover, they were writing in the pre-Heyerdahl era. For in earlier months of the same year as took the United Nations secretariat to Mexico, Thor Heyerdahl and a number of companions sailed from the Peruvian coast to Raroia, one of the most northerly islands of the Tuamotu archipelago, not far from Tahiti. Their vessel was a sailing-raft of balsa logs built as nearly as possible like those of the ancient Peruvians, and the journey was safely accomplished by making use of the appropriate currents and winds. With Heyerdahl's main theory of the origin of the Polynesians we are not concerned here, but would only mention that even his strongest opponents are at one with him in admitting that the Pacific could be navigated in sailing-rafts, to say nothing of more conventional non-European vessels. For it is quite clear that what could be done by balsa rafts, with sails and centre-boards, from east to west along latitudes 0° to 25 °S, could also be done by sailing-rafts of the South Chinese and Annamese pattern navigating (or drifting) from west to east along latitudes 25° to 45 °N. There are appropriate currents and winds in the winter and early spring to assist achieving this. They would be helped too by the climate of the North Pacific at these times, exceptionally warm for these latitudes.

It is interesting to ask how far back knowledge of eastward currents can be traced in Chinese literature. The Chinese name for the great Gulf Stream perpetually flowing north-eastwards was Wei-Lü, and it seems to have been known as early as the Warring States period. In the Chuang Tzu [Zhuang Zi] (The Book of Master Zhuang) of 290 BC we read:

> Of all the waters under Heaven there is none so great as the ocean.
> A myriad rivers flow into it without ceasing, yet it is never full, and
> the Wei-Lü (current) carries it continually away, yet it is never
> empty.

Wei-Lü might be translated 'the ultimate drain' or even 'cosmic sewer', and another term for it was Wu-Chiao [Wu-Qiao], 'converging and pouring away'. People sometimes said that it was a great rock, with an abyss or whirlpool into which the sea evermore discharged itself. In AD 1607, Ssuma Kuang [Sima

Guang] was quite·sure that the Fu-Sang country was to the west of the Wei-Lü current, and by 1744 there was a centuries-long tradition that the Wei-Lü was the current now known as the Kuroshio, which is one that Chinese mariners voyaging to America would have used.

Possibly the fable of the abyss discouraged now and then sailors of the Chinese coasts from launching forth on to the main – very similar stories plagued the mariners of Europe. A search in the maritime literature of old China would doubtless reveal much about the current as well as the belief in the maelstrom. For instance in the *Ling Wai Tai Ta* [*Ling Wai Dai Da*] (Information of What is Beyond the Passes) of 1178, Chou Chhü-Fei [Zhou Qu-Fei] says not a little about all this, but we must be content with a very brief extract: 'East of Shê-Pho [She-Po] is the Great Eastern Ocean Sea where the waters gradually begin to slope downwards.... Still further east is the place where the Wei-Lü drains into the world from which men do not return.' The statement about the point of origin of the Kuroshio current was right enough, though we should say the Philippines instead of Java; perhaps the 'bourne from whence no traveller returns' was the American continent rather than the abyss. Indeed, other and earlier accounts that no one had ever crossed the Pacific may have arisen because those who had tried to do so on primitive craft never returned. They had very little chance to do so, since before relatively modern times no general understanding of the régime of the eastward winds and currents could develop. Of those who made the journey, within the span of the first millennia BC and AD, many were probably fishermen or traders, involuntary carriers of culture to the Americas. Assuredly sometimes the great voyage was undertaken on purpose for one reason or another, though not with knowledge of any land at the other end.

Surprise that the Chinese did not explore the Pacific can only be felt by those unacquainted with Chinese literature. Indeed, all our considerations so far, cast a rather new light on the statements.in ancient Chinese writings about voyages into the Pacific. To be brief we must concentrate attention on the Chhin [Qin] period, when the rulers of China were convinced that drug-plants giving longevity or immortality were to be found on islands in the Eastern Ocean. In the late third century BC many sea-captains were sent out in search of these, generally without success. Though the only name that has come down to us is that of Hsü Fu [Xu Fu], the whole story of such activities is of such interest for the early maritime history of China that it is worth some examination. Ssuma Chhien [Sima Qian] refers to it four times in his great *Shih Chi* [*Shi Ji*], the oldest of the dynastic histories, finished in 90 BC, and his texts have much to tell us. First, in his chapter on Fêng [Feng] and Shan sacrifices, he says:

> From the time of the Kings of Chhi [Qi] (State) [378 to 279 BC], people were sent out into the ocean to search for (the islands of)

Phêng-Lai, Fang-Chang, and Ying-Chou [Peng-Lai, Fang-Zhang, and Ying-Zhou]. These three holy mountain (isles) were reported to be in the midst of Po-Hai [Bo-Hai] (the Gulf Pei-chih-li [Bei-zhi-li] in the Yellow Sea), not so distant from human (habitations), but the difficulty was that when they were almost reached, boats were blown away from them by the wind. Perhaps some succeeded in reaching (these islands). (At any rate, according to report) many immortals live there, and the drug which will prevent death is found there.... Before you have reached them, from a distance they look like clouds, but (it is said that) when you approach them, these three holy mountain (isles) sink down below the water – or else a wind suddenly drives the ship away from them. So no one can really reach them. Yet none of the lords of this age would not be delighted to go there.

The text now goes on to tell how the emperor Chhin Shih Huang Ti [Qin Shi Huang Di] despatched a crew of young men and girls to find these islands, and how they could not reach them because of the contrary winds. So far, of course, the whole atmosphere is legendary and magical. But more substantial facts follow. In 219 BC, when the emperor was on one of his periodical excursions along the eastern coasts, we are told that commemorative tablets were set up on the shore at Lang-ya and elsewhere. After this had been done, Hsü Fu [Xu Fu] and others told the emperor about the three magic mountain islands, and said:

> ... 'We beg to be authorised to put to sea, after due purification, and accompanied by a (suitable number of) young men and girls, go forth in search of these islands'. (The emperor approved this petition and) despatched Hsü Fu with several thousand young men and maidens to go and look for (the abodes of) the immortals (hidden) in the Eastern Ocean.

One would give a good deal to know what kind of craft they sailed in, and it would not be a surprise to discover that whole fleets of sailing-rafts were employed. Whatever they were, they cost an enormous sum, and in 212 BC the emperor bitterly complained about the expenses, always fruitless, which were being incurred.

The best account of Hsü Fu's [Xu Fu's] actions is given by Ssuma Chhien [Sima Qian] in his biography of the Prince of Huai-Nan. There he says that on his return Hsü Fu began to make excuses:

> 'In the midst of the ocean I met (on an island) a great Mage who said to me "Are you the envoy of the emperor of the West?", to which I replied that I was. "What have you come for?" said he, and I

answered that I sought for those drugs which lengthen life and promote longevity. "The offerings of your Chinn [Qin] King", he said, "are but poor; you may see these drugs but you may not take them away". Then going south-east we came to Phêng-Lai [Peng-Lai], and I saw the gates of the Chih-Chhêng [Zhi-Cheng] Palace, in the front whereof there was a guardian of brazen hue and dragon form lighting the skies with his radiance. In this place I did obeisance to the Sea Mage twice, and asked him what offerings we should present to him. "Bring me young men", he said, "of good birth and breeding, together with apt virgins, and workmen of all the trades; then you will get your drugs".' Chhin Shih Huang Ti [Qin Shi Huang Di], very pleased, set three thousand young men and girls at Hsü Fu's disposal, gave him (ample supplies of) the seeds of the five grains, and artisans of every sort, after which (his fleet again) set sail. Hsüan Fu (must have) found some calm and fertile plain, with broad forests and rich marshes, where he made himself king – at any rate he never came back to China.

Thus Ssuma Chhien suggests that though Hsü Fu humoured the emperor's Taoist [Daoist] beliefs, he really knew that there were good and vacant lands away in the east, and planned to make off there. Later generations believed that he had settled in Japan, and a tomb shrine still exists at Shingu. But this has not the value of an independent tradition since Japanese scholars all through the ages were familiar with the *Shih Chi* [*Shi Ji*]. Archaeological evidence is more compelling, for Chinese influence reveals itself strongly in the artifacts of the Japanese Yayoi period (first centuries BC and AD). Nevertheless it may be almost equally likely that the story of Hsü Fu's disappearance conceals one voyage at least to the American continent. Where he and his people went we shall probably never know for certain. But what sails the settlers had, or what means they took to steer their vessels over the broad waters, are matters not at all beyond conjecture.

4

Navigation

THREE AGES OF PILOTING

We must turn now to the techniques of Chinese navigation. We have come close to the subject already when discussing the magnetic compass and the contrasts between Portuguese and Chinese long-distance navigation. But really to examine Chinese navigation, even briefly, we must first try to summarise in a few paragraphs what is known of navigation in the Western world. Only in this way can we set up some standards of comparison. In doing so it will be convenient if we distinguish three periods, those of (a) primitive ·navigation, (b) navigation using measurements of various kinds, and (c) mathematical navigation. The second period will begin about AD 1200 in the Mediterranean, and the beginning of the third at, or a little before, 1500.

The primitive period

It cannot be said that the mariners of primitive ages were without guidance from astronomy; from very early times they steered by the stars and the sun. By night they could orient themselves by the stars close to the north celestial pole and, as far as the Egyptians were concerned, also by the stars of the 36 constellations into which they divided the sky, the decans, so-called because each occupied 10° around the sky, i.e. 10° of longitude. They could also gain some idea of their position northwards or southwards – their latitude – by observing the height of the Pole Star seen against masts and rigging. By day they could watch for the rising and setting points of the sun and so construct a plan of geographical directions or wind-rose.

Time and distance estimation were still crude. It amounted to no more than a count of day and night watches together with a guess at the distance covered. Yet these ancient pilots were observant men, men who took soundings, examined samples from the sea bed, noted the prevailing winds and currents, and recorded in their early rutters or sailing instructions depths, anchorages, landmarks and tides. Nor did they forget to use the services of shore-sighting birds. But if they made any charts, not one has survived. The art of such a man was summed up in an Indian text from the fourth century AD, which says of the famous pilot Suparaga:

> He knew the courses of the stars and could always readily orient himself; he also had a deep knowledge of the value of signs, whether regular, accidental or abnormal, of good and bad weather. He distinguished the regions of the ocean by the fish, by the colour of the water, by the nature of the bottom, by the birds, the mountains (landmarks), and other indications.

Primitive these arts may have been, but we must not miss the significant point that such navigation was already oceanic.

Measurement in navigation

Measurement was the keynote of our second period, when pilots increasingly adopted new inventions and novel practices rather than relying on guesswork and the help of the gods. After about AD 1185 in the Mediterranean, the emphasis rapidly shifted from celestial to terrestrial observation, for the discovery of the magnetic compass now became known and used there, causing a veritable revolution in the art of seafaring. Not only could the way forward be known through days and nights of overcast cloud or storm, but also the quantitative accuracy of dial readings round the horizon (azimuth readings) brought many important developments in its train. Naturally the windrose became more complex, but in addition it transferred itself to parchment in the form of those 'portolan' charts with their interlaced 'rhumb-lines' (lines of equal compass bearing) radiating from a series of centre points. The earliest dated sample of such a chart is of 1311, but historians of map-making customarily date its elaborate counterpart, the Carta Pisana, at about 1275, and there is some evidence from texts that sea-charts of this type go back to 1270 when St Louis of France was on his way from Aigues-Mortes to the Crusades.

In the light of these new bearing-and-distance charts other practices needed improvement, notable among them time-and-distance measurement. Its improvement became vitally important, and there was much talk of 'orologues de mer' from about 1310 onwards, then 'dyolls' from 1411, and 'running-glasses' from about 1490; all were one and the same thing, namely the hour-glass or sand-glass, regularly turned (with a chant) by the petty officer of the watch. Moreover, the pilot might well be obliged by the winds, as of old, to follow a course rather different from what he intended, but from about 1300 trigonometry, a branch of mathematics new to Europe, made it possible for him to have traverse tables from which he could calculate very easily how much of his intended course he had made good after a certain time, and how far he would have to sail to get back on to it. Though we have no example of such tables before 1428, we do know that they were being developed as early as about 1290. Doubtless an absolute limiting factor for the appearance of these computing pilots of the late thirteenth and early fourteenth centuries was the popularisation of Hindu–Arabic numerals; this was complete just after the first use of the magnetic compass by Mediterranean sailors (late twelfth century), though the new figures had first become known to the West before the end of the tenth century. As for the compilation of rutters, our oldest example of this period is the Italian *Compasso da Navigare* of 1253, which describes the whole Mediterranean in terms of the bearing-and-distance system which the portolan charts were shortly to embody, together with much other information of value to pilots.

So far there is no dispute, but real difficulty arises when we come to the application of more precise measurement for guidance by sun and stars. We shall come to what happened in East and Sourth East Asia in due course, but we are here concerned chiefly with the problem of precisely when in the fifteenth century Portuguese navigators of the ocean seas began to measure the altitude of the Pole Star with accuracy, using either a simplified seaman's astrolable, or a simple quadrant.

In 1321 Levi ben Gerson gave the first description in the West of the use of the cross-staff and 'Jacob's Staff' and it has often been assumed that this marks its first Western use. This is mistaken. The Portuguese cross-staff or *balestiha* was probably not a direct development from this astronomical and survey instrument, but rather from the *kamal* (Fig. 210) of the Arabs which had been encountered in the Indian Ocean. For much evidence has been collected showing that the pilots of the Mediterranean, who were primarily concerned with east–west voyages, never took altitudes at all until a very late date. Our oldest existing seaman's astrolabe dates from 1555 and our oldest dated drawing pushes this back to 1525, though we may accept its use from 1480 onwards. The use of the sea-quadrant after 1480 is also acceptable enough, but any earlier is uncertain. The great services of the Portuguese in determining the régime of winds and currents in the Atlantic are questioned by no one, and it is agreed that by about 1480 they were making fairly precise measurements of the altitude of the Pole Star, because during the following years they were making latitude determinations along the whole length of the African coast. This coincided with the recovery of the method of charting parallels of latitude and 'climates' (bands of the Earth's surface lying between two parallels) devised originally by the Greek astronomer Ptolemy in the second century AD.

Mathematical navigation

We are now at the threshold of the Renaissance, which marks the transition from the second to the third period. From 1500 onwards new aids for sea pilots came tumbling out of the cornucopia of the 'new, or experimental philosophy' – the beginning of modern science – in a wealth almost as bewildering as that of our own computer age.

Beginning again with the astronomical field, first mathematical tables multiplied – there were noon solar tables (1485), Southern Cross tables (1505), solar rising and setting tables (1595), to mention only some – and these led in 1678 onwards to the collections of tables to be found in nautical almanacs. At the same time instruments themselves were continually being improved, the first vernier-type graduations (allowing easy and accurate reading of fractions of each division of the scale) arrived in 1542, the cross-staff was superseded by the more accurate Davis back-staff (1594), and then came the still more precise reflecting sextant and octant (1731). Of signal importance, too, was the invention of the marine chronometer which, though suggested in 1530, was not a practical proposition until about 1760; with its adoption the age-old problem of finding longitude at sea was solved. In 1700 the coming of the marine barometer meant that weather forecasting at sea was now possible.

During all this time, knowledge of the earth's magnetism increased. Declination having become known to Europeans in the last decades of the fifteenth century, its variation in different localities was first plotted in 1535. In 1699 Edmond Halley voyaged across the Atlantic and enlarged this knowledge, so vital for mariners, almost to a world scale. In addition, from 1500 onwards the Cardan suspension was taken to sea, and the compass swung in gimbals. At the same time great advances were made in the measurement of a ship's speed, the old rough rules for estimating it being replaced in 1574 by having a floating weight (the log) at the end of a knotted line, or the accurate timing by a sand-glass of an object in the water as it moved past two marks on the ship itself. Finally, there were advances in the written records. From 1500 onwards the experiences of Europeans in the Indian Ocean gave rise to ever more detailed rutters, while latitude charts with graduated meridians led directly to the Mercator map projection (1569) and thence to others. These allowed plots drawn on them to show latitude, longitude, bearing and course, approximately correctly. Improvements in terrestrial globes led to the method of sailing in 'great circles' as first explained by Pedro Nunes and many others from 1537 onwards.

In what follows, we shall see clearly enough that though Chinese pilots never by themselves entered the third of our phases, mathematical navigation, they had sailed into the second phase – careful measurement in navigation – some two or three centuries ahead of the Europeans. They too deserve their praise:

If Pilots painful toil be lifted then aloft
For using of his art according to his kind,

What fame is due to them that first this art outsort,
And first instructions gave to them that were but blind.

STAR, COMPASS AND RUTTER IN THE EASTERN SEAS

It is hardly to be questioned that from the earliest times when Chinese shipmasters sailed their vessels out of sight of landmarks, they steered by the stars and the sun. Chang Hêng [Zhang Heng] was probably referring to their starcraft when he wrote in his *Ling Hsien* [*Ling Xian*] (The Spiritual Constitution of the Universe) of AD 118: 'There are in all 2500 (greater) stars, not including those which the sea people [or sailors] observe.' This raises the ghost of a literature long lost, and now hard to interpret, but perhaps highly relevant. For the *Khai-Yuan Chan Ching* [*Kai-Yuan Zhan Jing*] (The Kai-Yuan reign-period Treatise on Astrology (and Astronomy)), of the eighth century, often quotes an ancient *Hai Chung Chan* [*Hai Zhong Zhan*] (Astrology (or Astronomy) of (the People) in the Midst of the Sea (or, of the Sailors)), the antiquity of which now seems certain, as we shall see.

The expression *hai chung* [*hai zhong*] appears in the title of six books given in a Han (first century AD) bibliography and it has three possible interpretations. These are (*a*) that it meant the people of some foreign countries or islands overseas, (*b*) the Chinese as opposed to overseas people, (*c*) the seafaring men of China's coastal provinces. Moreover, the material of the books could have been primarily astrological or primarily concerned with navigation. This first has been favoured by Western scholars (so often prone to doubt Chinese originality); the second was argued in the seventeenth century by a Chinese scholar, though not very convincingly; and the third seems now the most reasonable. It was also thought to be the correct interpretation by the great thirteenth-century scholar Wang Ying-Lin. Certainly the expression had this connotation in the eighth century, and we shall not go far wrong if we identify the Hai Chung corpus of books as the work of those 'magicians' of the Warring States period and early Han (third century BC) who lived along the eastern coasts of the North China Plain, the 'mathematical practitioners' of the earliest stages of Chinese navigation. Their skills were doubtless undifferentiated, and it would be impossible to disentangle in them the components which today we should call astrology, astronomy, stellar navigation, weather-prediction, and the lore of winds, currents and landfalls; all the more so since these elements were still wholly confused down to the end of the seventeenth century AD in Europe.

At all events we can now form some idea what kind of men those 'sea-going magician-technicians' were whom the emperor Chih Shih Huang Ti [Zhi Shi Huang Di] interrogated in 215 BC, nameless though they are. However, one certain figure from among these magicians emerges in Wang

Chung [Wang Zhong]. The *Hou Han Shu* (History of the Later Han Dynasty) of the fifth century AD tells us that Wang Chung, who lived in the coastal province of Shantung [Shandong], 'delighted in Taoist [Daoist] techniques and was well versed in astronomy', so when trouble came during the rebellion of the Lü family about 180 BC, he put to sea with all his people and sailed eastward to Lo-lang in Korea, where he settled in the mountains.

Chinese pilots in the period of primitive navigation made use also of all those ancient aids which have already been mentioned. But it was they who brought this period to an end by being the first to employ the magnetic compass at sea. This great revolution of the sailor's art, which ushered in the era of true measurement in navigation, is solidly attested for Chinese ships by AD 1090, just about a century before its initial appearance in the West, as we saw in chapter 1. Incidentally, the first Chinese text which mentions this also makes reference to astronomical navigation and soundings, together with the study of sea-bottom samples. In the twelfth century, two further Chinese accounts followed before the first European mention, each emphasising the value of the compass on cloudy and stormy nights. Admittedly, the precise date at which the magnetic compass first became the mariner's compass, after a long career ashore with the geomancers, is not known, but some time in the ninth or tenth century would be a very probable guess. We say this because these centuries saw rather refined measurements of magnetic declination which could only have been made by the use of the needle. Before the end of the thirteenth century (Marco Polo's time) we have compass bearings recorded in print, and in the following century, before the end of the Yuan dynasty, compilations of these had begun to be produced.

As we also saw in the first chapter, the earliest description we have of a floating compass using small pieces of magnetised iron or steel dates from just before 1044; it involves a thin leaf of the material cut into the shape of a fish. To floating compasses of one kind or another Chinese sailors remained faithful for nearly a millennium, though the Dutch influence of the sixteenth century resulted in the adoption of a dry suspension in Chinese vessels. But it must also be remembered that the Chinese compass-makers employed a very delicate form of suspension which automatically compensated for variations in dip, and still impressed Western observers as late as the beginning of the nineteenth century.

Chêng Ho's [Zheng Ho's] remarkable series of expeditions between 1400 and 1433 must always remain a focal point in the history of Chinese navigation. By great good fortune, certain portolan-style maps dating from about this period, tracing the routes followed by these and other Chinese ships and convoys, have been preserved intact. Early in the seventeenth century they were printed as the last chapter of the important *Wu Pei Chih* [*Wu Bei Zhi*] (Treatise on Armament Technology). These charts are extremely dis-

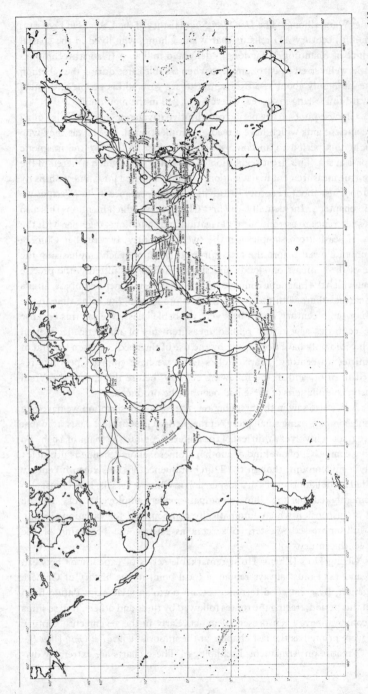

Fig. 207. Comparative map of the Chinese and Portugese discoveries and navigations in the fifteenth century. Chinese dates are bracketed if drawn from textual evidence before the fifteenth century; dates given for that period indicate at least the earliest visit recorded.

167

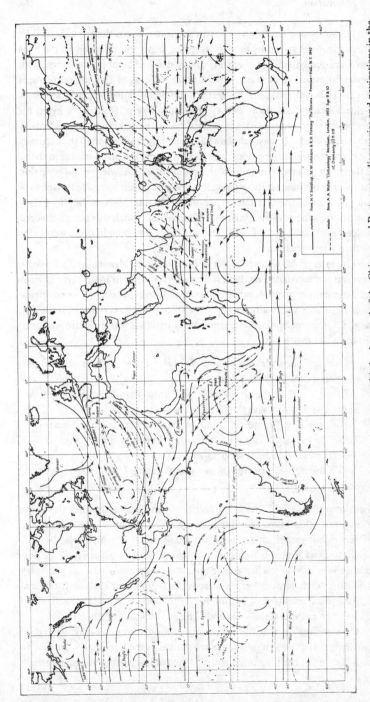

Fig. 208. Map to illustrate the meteorological and oceanographic background of the Chinese and Portuguese discoveries and navigations in the fifteenth century.

torted but schematic, and ships' courses are drawn across their oceans like the tracks in the maps issued by modern shipping companies. However, the lines of travel are accompanied by legends giving detailed compass-bearings, with distances in numbers of watches, and notes of most of the coastal features which could be important in navigation. These notes include indications of half-tide rocks and shoals as well as ports and havens. Routes are given for inner and outer passages of islands, sometimes with preferences if outward- or homeward-bound. Modern scholars have paid much attention to the accuracy of the diagrams and descriptions, and to identifying the place-names in them, with the result that there is now a high opinion of the knowledge and precision of the records of these Chinese navigators. We may gain some idea of the skill of the pilots from the fact that in circumnavigating Malaya they laid their course through the present Singapore Main Strait, which was not discovered (or at least not used by) the Portuguese until they had been in those waters for more than a hundred years.

The interest of the last chapter of the *Treatise on Armament Technology*, however, is not exhausted by these schematic charts. Four instructive navigational diagrams are given, summarising star positions to be maintained during as many regular voyages. Figure 209 shows the pilot's directions for travelling between Ceylon and Sumatra, and the notes surrounding the picture lead us to the heart of the matter:

[Above] The Pole Star to be 1 *chih* [*zhi*] above the horizon, and the 'Imperial canopy' 8 *chih*.

[To the left] In the north-west the Pu ssú [Bu si] stars to be 4 *chih* above the horizon; and the same in the south-west.

[Below] The 'Frame' or 'Bone' of the 'Lantern' (i.e. the Southern Cross) to be 14½ *chih* above the horizon. The twin stars of the 'Southern gate' to be level at 15 *chih*.

[To the right] in the north-east the 'Weaving Girl' to be 11 *chih* above the horizon.

The explanation of all this lies in the fact that in measuring the altitudes of the Pole Star and other stars, the pilots did not use the degrees adopted by the astronomers but rather 'finger-breadths' (*chih*), each of which was divided, into eight or more probably four parts. Moreover, for this voyage the Pole Star was either very low on the horizon or invisible, and it was therefore necessary to substitute for it some other stars from the region near the north celestial pole. (They used stars from the Chinese constellation Hua kai [Hua gai] which comprises parts of our Cassiopeia and Camelopardus constellations.)

Once we realise that the navigators of the China Seas and the Indian Ocean depended quite as much on polar altitudes as the Portuguese came to do towards the end of the fifteenth century, a host of fascinating questions arises.

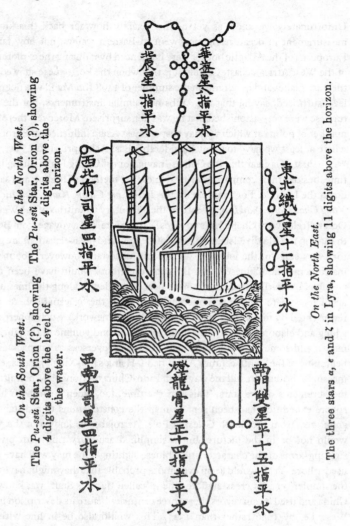

辰星一指平水　華盖星八指平水

西北布司星四指平水

西南布司星四指平水

東北織女星十一指平水

燈龍骨星正十四指半平水　南門雙星平十五指平水

On the North West.
The *Pu-ssŭ* Star, Orion (?), showing 4 digits above the horizon.

On the South West.
The *Pu-ssŭ* Star, Orion (?), showing 4 digits above the level of the water.

On the North East.
The three stars α, ε and ζ in Lyra, showing 11 digits above the horizon.

On the South.

The Southern Cross showing 14½ digits above the horizon, and the *Nan-mên-shuang-hsing* (Centaurus, α and β) showing 15 digits above the horizon.

Fig. 209. One of the seventeenth century Chinese navigational diagrams of the *Treatise on Armament Technology*, that for the Ceylon-Sumatra run, reproduced from a book by G. Phillips.

Unfortunately we know as yet neither exactly how far back this kind of measurement in ocean navigation went in Eastern waters, nor how far the Europeans of the Atlantic border were influenced by it during the explorations of the West African coast. Certain it is that when the Portuguese showed him their astrolabes and quadrants in the summer of 1498 Ibn Majid was not in the least surprised, saying that the Arabs had similar instruments, but the Portuguese were very astonished that he was not surprised. Moreover, there are a number of points at which we may suspect East Asian influence on Europe, or where at least we have to grant considerable East Asian priority.

First, it is clear that the Chinese navigators of Chêng Ho's [Zheng Ho's] time, besides their compass-bearings, knew the method of finding and sailing down the latitude. For example, in the *Hsi-Yang Chhao Kung Tien Lu* [*Xi-Yang Chao Gong Dian Lu*] (Record of the Tributary Countries of the Western Oceans [relative to Chêng Ho's voyages]), there is talk of a voyage from Bengal to Malé in the Maldive Islands by way of Ceylon, and polar elevations are given for every stage of the journey. We are still uncertain, however, about what instruments the Chinese used. By 1400 quadrants would have been quite possible, but earlier there might have been some device along the lines of the armillary sphere (where the reference circles on the 'celestial sphere' were represented by metal rings fixed to a skeleton framework). Such spheres had a long and elaborate history in China (see the second volume of this abridgement), and some such apparatus had been used overseas as far back as the beginning of the eighth century AD, when I-Hsing's [Yi-Xing's] teams sent to measure meridian arcs all the way from Indo-China to Mongolia, took altitude measurements of the stars. This was the time, too, when a southern hemisphere expedition had been sent to map the constellations to a distance of about 20° from the South Celestial Pole. Astrolabes as known in the West would not be in the picture, but a simplified armillary ring with pivoted sighting-arms or the characteristic Chinese sighting-tube may well have been used ashore. Yet it would seem even more probable that they might have used the simpler types of cross-staff, for the so-called 'Jacob's Staff' was known in China and used by surveyors there three centuries before its description in the West, i.e. by 1086 rather than 1321. This would also be in line with the practices of Arab and Indian pilots.

The problem of maritime charts is also very obscure. That they existed is implied by many Chinese texts, but the only ones to have survived are the diagrams in the *Treatise on Armament Technology*. Nevertheless, the tradition of careful map measurement was much stronger in China than it was in Europe, so that already by 1137 a superb map on a scale of 100 *li* to the division could be produced, and there is little reason to think that the much larger map of Chia Tan [Jia Dan] on a similar scale in 801 was any less good. Indeed, the principle of using a rectangular grid for mapping went back to the third century AD, and never gave place, as map measurement or quantitative cartography did in Europe, to the fancies of the wheel maps of religious cartographers. It is therefore interesting that in the work of Shen Kua [Shen Gua] late in the eleventh century we do have a hint that the grid was combined with lines showing compass-bearings (rhumb-lines), just as occurred two or three centuries later in the Mediterranean. But Shen Kua's work was terrestrial, not nautical, and it did not survive, though it is interesting to note that the Arabs had rectangular grid charts, often graduated in finger-breadths, and sometimes possibly in the Chinese fourteenth century 'Mongolian style' with names of ports and places but little geographical detail.

Lastly, there is the projection devised by Gerard Mercator in 1569; this was a great advance, but he never knew that he had been preceded by Su Sung [Su Song] five centuries earlier in a celestial atlas. In this the hour-circles between the lunar mansions formed the meridians, with the stars marked on each side of the equator according to their distances from the North Celestial Pole, thus giving a map projection in which small areas retained their correct shapes (a quasi-orthomorphic cylindrical projection). With such a brilliant background we may hope that archaeological dicoveries will reveal what charts were used by the master-mariners of the Sung [Song], Yuan and Ming.

As we have now seen, the magnetic compass, the portolan chart, the sand-glass and the use of traverse tables formed a closely connected knot of contemporary techniques. Little can be said of traverse tables because these have not so far been recognised in Chinese rutters, but the use of sand for time-measuring opens up curious perspectives. A memoir of 1951 on the history of the mariner's compass concluded that the sand-glass was not known or used on Chinese ships until the end of the sixteenth century when they acquired it from the Dutch or the Portuguese. But since that time much information has come to light about an important development in the history of Chinese mechanical clockwork which occurred about 1370. This was the substitution of sand for water as the means of driving mechanical clocks. Whether, when this happened, the Chinese continued with the link-work escapement of their clocks, or adopted gearing instead, remains unclear. Nevertheless, it certainly involved something new for Chinese clockwork, as well as for the more recent clocks of the West, which were acquiring it too; this was the use of a stationary dial with a moving pointer.

This new look is associated with the name of Chan Hsi-Yuan [Zhan Xi-Yuan], and there is no reason why one or more of his clocks should not have been carried (as the old water-wheel ones could not) on each of the great ships of Chêng Ho's [Zheng Ho's] fleet. In any case it is clear that time-keeping by sand-flow was very much in the minds of the Chinese of that time. It is necessary therefore to re-examine the Western traditions which make the sand-glass begin with Liutprand of Cremona, the tenth-century Italian diplomat, historian and bishop, and reconsider the suggestion made in the early years of this century and long ignored, that the hour-glass came to Europe from the East.

To begin with, a modern Chinese historian has argued to considerable effect that since nautical watches are mentioned or implied in many descriptions of Chinese navigation from the beginning of the twelfth century onwards, the measurement of such units must have necessitated the sand-glass, since no water-clock would have been imaginable at sea. Again, a Western nautical historian claims to have traced the Western nautical sand-glass back to the famous Venetian glass industry of the late twelfth century. Thus the possibility presents itself that together with the magnetic compass and the stern-post rudder, it may have formed part of one of those transmissions from Asia which we find in so many fields of applied science. But against this there is a serious argument and another way out.

The sand-glass implies blown glass, and it has been found that the glass-blowing art appears to be wholly European and Western, though by no means glass-making itself. Is not the time-keeping 'joss-stick' the real answer to Chinese historian just mentioned? Burning incense in stick-like form is a practice which goes far back into China's Middle Ages, and it would have been very easy to measure time approximately enough with the 'joss-sticks' that were kept alight in the ship's shrine where the compass lay also. In this case the use of 'combustion clocks' at sea for watch-keeping gave a very practical and reliable 'proto-chronometer'. Yet the incense stick was so characteristic of Chinese religion and culture that it may have been difficult for it to spread to mariners of other cultures even though they might have found it very useful.

Let us return now to the altitude measurements in *chih* and *chio* [*zhi* and *jue*]. The remarkable feature of this system is that it was practically identical with that in use among the Arab shipmasters of the Indian Ocean, who expressed altitudes in the finger-breadth or inch. The system was long known to Europeans, mainly from the *Muhit* (The Ocean), a compendium of nautical instructions put together by the scholarly Turkish admiral Ibn Husain when staying at Ahmedabad in India in 1553 on his epic journey home after the destruction of his fleet. Later his chief sources became known, an Arab treatise of 1511 and a still earlier book of 1475 written by the Arab pilot who joined Vasco da Gama at Malindi in 1498. We know now that the Portuguese navigators made use of the system for some time afterwards. It is clear, then, that this tradition must have been in full use at the time of Chêng Ho's [Zheng Ho's] voyages. Moreover, when the measurements in the *Muhit* and its sources are compared with those in the Chinese *Treatise on Armament Technology*, they are found, generally speaking, to be in good agreement. The chief difference between the Arab and Chinese systems seems to be that when a 'substitute' polar mark-point was needed in equatorial latitudes, the Arabs chose the classical 'Guards' (beta and gamma Ursae Minoris) which they called the 'Calves', while the Chinese chose stars from the Hua kai [Hua gai] constellation. The north polar distances of these stars are very similar, though they lie on opposite sides of the Pole Star. Both Arabs and Chinese took an elevation of the Pole Star of one finger-breadth as the point at which its use was no longer reliable; they then changed from the Pole Star to their circumpolar stars, the Arabs taking eight finger-breadths to the 'Calves' and the Chinese eight finger-breadths to Hua kai as equivalent to one finger-breadth of the Pole Star.

The first Europeans who visited the southern hemisphere found it very strange that the northern Pole Star disappeared from sight. Marco Polo lost it in Sumatra on his voyage home in 1292 and recovered it at Cape Comorin (lat. 8 °N); Odoric of Pordennone remarked on the same thing some 20 years later. Though these and other Western writers did not record the southern stars, they were greatly impressed at the navigational use made of them, so much so indeed that they gave the impression that the magnetic compass was not used in those waters. Indeed Fra Mauro writes on his map that the Chinese junks '... navigate without a compass because they carry an astronomer who stands alone on the high (poop) and commands the navigation with an astrolabe', though it is probably the *kamal* not the astrolabe that he really meant.

The impression given about the compass by such writers was undoubtedly wrong. It seems likely to have arisen because the Mediterranean pilot of the fourteenth century never took his eyes off the needle, and gave orders to the helmsman accordingly while working out his course by bearing and distance. For the Asian pilot the compass was only one of his instruments, and the determination of position by star (and possibly even sun) sights was at least equally important. This was no doubt because the region sailed by the seamen of the Arabic tradition was one of relatively scanty, or at least very seasonal, rainfall, and frequent clear skies, so that orientation by the stars was more inviting and capable of more precision. At the same time their oceanic domain included both northern and southern hemispheres, with all that implied of astronomical challenge. And as the interruption of overcast skies was less, there was not the same reason to wax enthusiastic about the leading of the lodestone. It is true that its only begetters, the northern Chinese, had done so, but their words were enclosed in an ideographic language, not to be understood or appreciated by Westerners until comparatively modern times.

The question may be raised as to the mutual influence of the Arabic and Chinese navigators, but at present we hardly know enough to answer it. They had certainly been in contact before AD 1400, and measurements of altitude were certainly prominent in Arabic astronomy. On the other hand circumpolar mark-points for invisible stars were rather characteristically Chinese. Again though measuring in finger-breadths is not common in early Chinese

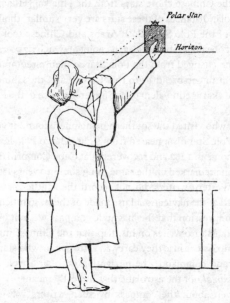

Fig. 210. The *kamal* in use for taking star altitudes, a drawing made by H. Congreve in 1850.

printed texts, they may still have been in use by pilots, for the rutters were hand-written, and certainly Chinese mention of land-surveying measurements using them is a good deal earlier than anything similar in Arabic culture. But, of course, the method could easily have arisen independently in both cultures.

What instruments were used by the Chinese pilots of the Yuan and Ming for taking star altitudes was long a puzzle, but of those employed among the Arab sailors a good deal was known. They were all forms of the cross-staff, including one, the tablet or *kamal* (Fig. 210), in which the wooden supporting rod was replaced by a knotted string. Supposedly earlier was a set of nine square boards or plates extended on a string or rod of standard length (normally the stretch of a man's arm). These devices measured the angle between the star and the horizon, not between the star and the zenith, which was much more difficult on shipboard. As we have seen earlier, there is evidence of the existence of the cross-staff in China in the eleventh century, 300 years before its traditional invention in Provence. It remains extremely probable, therefore, that Chinese pilots of the fifteenth century used some form of cross-staff.

That they used a version of the *kamal* as one kind of cross-staff has now been proved by the modern Chinese scholar Yen Tun-Chieh [Yen Dun-Jie]. He has brilliantly interpreted a passage in the *Chieh An Lao Jen Man Pi* [*Jie An Lao Ren Man Bi*] (An Abundance of Jottings by Old Mr (Li) Chieh-An [Jie-An]) written by Li Hsü [Li Xu], and which was printed in 1606. It runs as follows:

> The set of 'guiding star stretch-boards' ... has twelve plates in all, made of ebony, ranging gradually from small to large. The largest is more than 17.5 centimetres square. They are labelled 'one *chih* [zhi]', 'two *chih*' etc., up to 'twelve *chih*', all marked in fine script upon them; and they differ regularly just as a centimetre is divided into millimetres. There is also one ivory piece five centimetres square, and cut off at the corners so that it indicates half a *chih* (i.e. two *chio* [*jue*]), half a *chio*, one *chio* and three *chio*. This may be turned on one side or another facing you (in conjunction with one of the larger plates), and these lengths must be the measurements (required for right-angle triangle calculations according to the methods of the) *Chou Pei* [*Zhou Bei*] (*Suan Ching* [*Suan Jing*]) (Arithmetical Classic of the Gnomon and the Circular Paths of Heaven).

Evidently we have here a set of standard ebony tablets held at a fixed distance from the eye, not the single one with its knotted string that constituted the typical *kamal*. What is more, there is the interesting addition of a 'fine adjustment' in the shape of an ivory tablet with corners cut to standard edge lengths, held up at the same time to allow the measurement of fractions of a *chih*. Calculations by Yen Tun-Chieh [Yen Dun-Jie] showed to what angles these tablets corresponded, and from these it is clear that Chinese pilots at this time at any rate had four *chio* to a *chih*, not eight, though the half-*chio* was marked on the ivory fine adjustment plate.

How long before his own time this system had been in use, Li Hsü [Li Xu] does not say, but for many reasons we can be sure that Chinese pilots were using his method in the fifteenth century, and they may well have been doing so in the fourteenth or even the thirteenth. Evidence indeed seems to be growing that they were taking altitudes by the beginning of the twelfth. A text of 1124 suggests this, and strange confirmation comes from the *Sung Hui Yao Kao* [*Song Hui Yao Gao*] (Drafts for the History of the Administrative Statutes of the Sung [Song] Dynasty) compiled in the Sung. Here we read 'These (sea-going) ships should each be equipped with a 'Dipper-observer' (*wang tou* [*wang dou*]), ...'. The obvious meaning of *wang tou* is the long hollow tube (the 'sighting-tube') for determining the positions and altitudes of the stars of the Plough or Dipper constellation. But it might equally well have been a cross-staff or *kamal*. Perhaps therefore the measurement of stellar altitudes followed closely upon the measurement of azimuth directions by the Chinese pilots.

Summing up the present state of our knowledge about the development of measurement in navigation in the Eastern Seas, we have to start with the introduction of the mariner's compass on Chinese ships some time before 1050, possibly as early as 850. How soon this spread to the Indian Ocean we still do not know. Before 1300 there is hardly any evidence of taking star altitudes at sea by instrument, whether among Arabic or Indian pilots, and only very little for the Chinese navigators. But the pilots' book *Shun Fêng Hsiang Sung* [*Shun Feng Xiang Song*] (Fair Winds for Escort) tells us that from 1403 'the drawings of the guiding stars were compared and corrected', which suggests a considerable previous development during the fourteenth century. Broadly speaking, therfore, we may not be far off the truth if we say that when Ibn Majid met Vasco da Gama at Malindi, full measurements in navigation were some two or three centuries old 'East of Suez' but hardly one century old in the West.

Star altitude measurements continued to interest Chinese groups concerned with the navigation of traditional craft down to our own times. However, it seems that Jesuit influence can, possibly, be detected in the seventeenth century, with the use by the Chinese of 360° in a circle instead of their customary 365¼. But there is one strange puzzle. A diagram using the 360° and illustrating a 'Plan of North Polar Altitudes', was published in the 1715 edition of the *Ting-Hai Thing Chih* [*Ding-Hai Ting Zhi*] (Local Gazetteer of the Sub-Prefecture of Ting-Hai [Ding-Hai]); what is so interesting is that it contains a whole family of 16 quarter-ellipses, equally spaced, in that part which represents the visible sky. Similar, but much simpler, diagrams were used to demonstrate navigational astronomy in sixteenth-century England. This raises some difficult questions.

One of the problems of the astrolabe which could be used for determining stellar altitudes at sea was that as originally designed, it needed a different engraved plate (*tympanum*) under the open fretwork star-map (the *rete*) for every latitude. One answer was to use the Rojas projection, a map projection on the tympanum that was 'orthographic', i.e. in which the sighting-point for drawing the earth's globe lies at an infinite distance outside it. The result of this is to give parallels of latitude which become straight lines at the equator, and meridians which appear as semi-ellipses. The intervals between both parallels and meridians become less the further they are away from the centre of the whole. Such unequal intervals can clearly be seen in Jesuit diagrams and must have been intended in the Ting-Hai [Ding-Hai] one.

Actually the sixteenth-century astronomer Juan de Rojas Sarmiento never claimed the invention of the Rojas projection. Simpler forms of it had been sketched in antiquity, notably by the engineer Vitruvius (about 30 BC). Textual evidence suggests that the Islamic geographer al-Biruni thought of the same thing about AD 998, though his explanation needs further study, while several instruments inscribed with the same orthogonal projection but antedating Rojas by half a century or more, have come down to us.

Of course the Rojas projection was not the only one to suit every latitude. Best known and most commonly found is that of Ibn Khalaf, a Toledan astronomer who was working in 1040. Later in the same century (*c.* 1070) it was improved by the noted astronomer al-Zarqali, famous for the influential Toledan Tables of celestial positions of sun, moon, stars, and planets. This last was illustrated and minutely described in the well-known *Libros del Saber de Astronomia* of Alfonso X, King of Castile, produced about 1276, which has a strange connection with the 16 quarter-ellipses of the Ting-Hai [Ding-Hai] diagram. The front of the second instrument described in the *Libros del Saber* is engraved with al-Zarqali's improved scheme, but the back bears a diagram which has not yet been adequately explained. While one quadrant is ruled with lines giving the sines of the angles of the scale of degrees, the three others contain a series of semi- or quarter-ellipses. Because of their spacing, these ellipses cannot represent meridians on an orthogonal projection.

The *Libros del Saber* is far from being alone in giving this construction, for we find it also on earlier Muslim astrolabes, one of 1212 and another of 1252. Hence the question may be raised whether the equally spaced quarter-ellipses of the Ting-Hai [Ding-Hai] diagram do not derive from earlier direct contacts between Chinese and Arabic astronomical navigators rather than from later Jesuit intermediaries.

Before taking leave of the Chinese pilots it may be of interest to glance at the contents of two or three typical rutters. The first of these is the *Shun Fêng Hsiang Sung* [*Shun Feng Xiang Song*] (Fair Winds for Escort), composed anonymously about 1430 or at the close of the period of Chêng Ho's [Zheng Ho's] expeditions. Another is the *Tung Hsi Yang Khao* [*Dong Xi Yang Kao*] (Studies on the Oceans, East and West), compiled by Chang Hsieh [Zhang Xie] in 1618, a few years after Emmanuel Diaz had produced his explanation of the celestial sphere, but showing no evidence of western influences. Chang Hsieh was much more scholarly as a historian and geographer than the fifteenth-century sea-captain, but also seems to have had some personal experience of the sea.

Looking first at what the two texts have in common, we find they give abundant information on landmarks and general sailing directions with compass-bearings, and soundings in fathoms. Both include information on various destinations, though the anonymous compiler takes us further afield. Each contains tables of monthly and seasonal winds with copious advice on weather signs, observing the shapes of clouds, the behaviour of wind and rain, together with other meteorological events such as haloes round the sun. Tide tables are given, to which are added other signs such as the colour of the water and objects likely to be floating on it. Both supply the master with liturgical instructions, the earlier rutter being more concerned with patron saints and tutelary deities of the compass, while Chang Hsieh [Zhang Xie] likes to speak of Thien-Fei [Tian-Fei], the Mistress of the Heavens, a sailor's goddess who had the devotion of Chêng Ho [Zheng Ho] and his fleets.

The material which is only in the fifteenth-century text has some features of special interest. We hear of the method of selection of water for the floating compass, and of the proper way to make the needle float on it. Besides three tables of the 24 azimuth points, there is one listing only 14 of them as 'palaces of the heavens', and another associating them with the winds. Most interesting with regard to possible Arab influence is a small table entitled 'Principles of Star Observations', which gives the azimuth rising and setting points of four constellations, the Plough (Dipper), the Hua kai [Hua gai] constellation (significantly) (see page 168), the Southern Cross, and a star which is possibly Canopus. Now such points of rising and setting were the elements of which the Arab sailor's azimuth graduation of the circle was wholly constructed. Azimuth rising points of sun and moon throughout successive months are listed, together with corresponding water-clock divisions and lengths of night and day. The fact that mnemonic verses on this, and on the moon's times of rising and setting, follow suggests that the pilots were accustomed to pay attention to these data. The anonymous compiler also gives mnemonic verses about lightning as a weather-sign. Lastly he adds something about the determination of currents and tides, and the calculation of watches. It is here that we find mention of the floating wood method of logging the speed of the ship.

Another anonymous rutter entitled *Chih Nan Chêng Fa* [*Zhi Nan Zheng Fa*] (General Compass-Bearing Sailing Directions) exists as a manuscript appended to a military encyclopaedia, the preface of which is dated 1669. Besides much weather-lore and rhumb bearings not yet analysed, it has a subsection on star sights with diagrams of the constellations. Observations of the stars of the lunar mansions given would provide not only the time of night but also the latitude by a simple calculation. It is very probable therefore that further researches in the literature will bring to light nautical tables marrying the greatest heights reached by the lunar mansion constellations with the positions of circumpolar stars lying below the horizon, similar to the Arabic lists.

Lastly, a word on tide tables. Since several of the exisiting Chinese rutters include forms of these, it is worth recalling that sea tides were studied in China earlier than in Europe. Authoritative histories still inform us that the oldest tide-table for a particular port is the early thirteenth-century 'fflod at london brigge', but it is now clear that Yen Su's *Hai Chhao Thu Lun* [*Hai Chao Tu Lun*] (Illustrated Discourse on the Tides) of 1026 contained a detailed tide-table for Ningpo [Ningbo]. Not much later, in 1056, Lü Chhang-Ming [Lu Chang-Ming] drew up a tide-table for Hangchow [Hangzhou] which was inscribed on the walls of a pavilion on the banks of the Chhien-thang [Qian-tang] River. The Chinese pilots of the Yuan to the Chhing [Qing] had thus a great tradition behind them.

Terrestrial globes

If we wanted to find a terrestrial globe at sea on board a modern liner there might be an ornamental one in the reading-room but we should certainly not expect to find one on the bridge. Yet there was a time, in the late sixteenth and early seventeenth centuries, when such objects figured prominently among navigational instruments. Yet it is very improbable that such a globe could have been found on a Chinese vessel, even on Chêng Ho's [Zheng Ho's] flagship, for representations of this kind were not in the Chinese tradition; or not exactly, for the statement is a half-truth and needs explaining.

First let us recall what happened in the West. The usual belief that the first maker of a terrestrial globe was the Stoic Crates of Mallos (about 160 BC) is authorised by the words of geographer Strabo (63 BC to AD 19). After Strabo little or nothing is heard of terrestrial globes in the West, but the tradition must have been transmitted to the Arabs for in 903 the Persian geographer Ibn Rustah gave a good description of the terrestrial as well as the celestial sphere. A few centuries later the Latins could do the same, as is shown by the Englishman Sacrobosco's *Tractatus de Sphaera* (about 1233), popular until the Renaissance. Now it was later in this same century that the astron-omer Jamal al-Din, heading a scientific co-operation mission form the Ilkhan of Persia to the court of China in 1267, took to Peking [Beijing] a terrestrial globe (or at least the design of one); 'a globe to be made of wood', says the *Yuan Shih* [*Yuan Shi*] (History of the Yuan (Mongol Dynasty), 'upon which seven parts of water are represented in green, three parts of land in white, with rivers, lakes, etc. Small squares are marked out so as to make it possible to reckon the sizes of the regions and the distances along roads.' But the idea did not catch on.

Why it did not is very hard to say, for there was much in Chinese tradition to welcome it. The great cosmologists of the Han repeatedly said that the earth floated in the heavens like the yolk in a hen's egg, or that the earth was 'as round as a crossbow bullet' suspended in space. When Hsiung Ming-Yü [Xiong Ming-Yu] 1500 years later illustrated his treatise on astronomy and geography with a pleasant picture of Chinese junks circumnavigating an upside-down ocean he significantly made use of the same phraseology.

Evidence is accumulating, moreover, that the belief in the spherical shape of the earth was much more widespread in medieval Chinese culture than has often been thought. And in fact the astronomers of China had actually been making terrestrial globes for centuries, but not on the same scale of size as their celestial ones, rather as quite small earth models held on a pin at the centre of their armillary spheres which were used for demonstration. What arrangement exactly the pioneer instruments of Châng Heng [Zhang Heng] (AD 125) and the astronomer Lu Chi [Lu Ji] (225) had is uncertain, but it is quite clear that a little later, in 260, the astronomer Ko Hêng [Ge Heng] placed a model earth inside his armillary. It is also certain that at about the same time Wang Fan, another astronomer, preferred to represent the earth horizon by sinking the sphere to the right extent within a flat-topped box. Ko Hêng's plan however was followed by many other instrument-makers down to the close of the sixth century, after which came the epoch of mechanical devices in which the earth was represented in a variety of ways. At least one of these, a Korean example from the eighteenth century, still probably exists, and its small inner earth model is marked with all the land-masses known to modern geography. Thus paradoxically although the terrestrial globe in its fully enlarged form was unfamiliar to Chinese culture, there had been earth models within armillary spheres from the third century AD onwards, a practice which did not begin in Europe until the end of the fifteenth century.

Two Chinese terrestrial globes still exist, but both of them are of Renaissance type. The first is a product of the Jesuit mission to China, a painted lacquer globe 57.5 centimetres in diameter, constructed in 1623 under the guidance of Emmanuel Diaz and Nicholas Longobardi. The other is a very different object, much smaller, having a diameter of a little under 30 centimetres, and made of silver sheet metal on which the map and inscriptions were incised before being covered with translucent cloisonné enamel in bright blue, green, violet and other colours. It carries no statement of origin, and internal evidence can only date it as of some time probably between 1650 and 1770; but a considerable similarity in place-names shows that it shares a common source with the world-map of the Chhing [Qing] cartographer Chuang Thing-Fu [Zhuang Ting-Fu], one edition at least of which was printed as late as 1800. The place-names on the globe are meant to be read with the South Pole uppermost. On it Australia is joined both to New Guinea and Tasmania; the East Indies are also badly drawn, for Borneo is placed between Malaya's tip and Java, a feature which suggests the absence of any Jesuit influence, and makes it more probably the product of some northern Chinese cartographer unfamiliar with the South China Sea. A large Antarctic continent (wrongly attached to New Zealand) is shown however. The drawing of California as an island might indicate a date around 1700, for the magnetic world-map of the astronomer Edmond Halley is one of the last to represent it in this way.

5

Propulsion

Introduction

Sails may be defined as pieces of textile fabric outstretched upon ships in various ways so that the pressure and flow of the wind can be utilised to drive the vessel upon its course. Sails, and combinations of sails, are of manifold shapes and arrangements (rigs), almost bewildering in their complexity, yet one single creative thread runs through them all – the desire of man to release himself from natural servitude by sailing, not only with favourable breezes but also directly into the eye of the wind. Though sails alone were never to permit him to accomplish this feat, there was a point of maximum efficiency which could be reached in sailing to windward, and that he ultimately attained. The history of this branch of nautical technology might thus be epigrammatically described as the advance from SW to NW by N in the face of a due north wind – a difference of nine points of the compass, yet one which took three millennia to achieve.

For the understanding of what follows it is necessary that we should bear in mind the chief varieties of what have been called the 'primary' types of sail. These are sketched in Fig. 211. In these diagrams the masts and supporting pieces (spars) are shown by a heavy line, while the free edges of the sail are left light. First, there is the square-sail, oldest and simplest, symmetrically hoisted, necessitating always an upper supporting cross-piece (a yard), but at different times and places with or without such a similar cross-piece at the bottom (a boom) (A, A'). The square-sail is the only principal sail which always receives the wind upon the same surfaces. With a wind from behind and to the right of a ship (i.e. aft from the starboard quarter), the right-hand side of the sail (the starboard side) would be braced forward while the left-hand side (the port side) would be aft of the mast. Then when the wind changed to the other, port, quarter, or when the ship changed course, the position of the sail would be reversed; in other words, the forward or weather yard-arm became

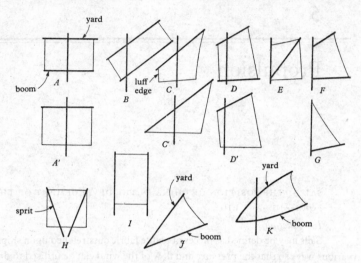

Fig. 211. Principal sail types. Masts, yards, booms, sprits, gaffs, etc., in thick lines; sail edges in thin. No attempt is made to indicate relative sizes of sails.

A   square-sail with boom
A'   square-sail without boom (loose-footed)
B   Indonesian canted 'square' (rectangular) sail, with boom
C   lateen sail with short luff (front) edge
C'   lateen sail without luff edge
D   lug-sail with considerable luff area forward of the mast
D'   lug-sail with reduced luff area forward of the mast
E   sprit-sail
F   gaff- or yacht-sail
G   leg-of-mutton sail
H   Indian Ocean bifid- or two-legged mast sprit-sail
I   Melanesian double-mast sprit-sail
J   Oceanic (Polynesian) sprit-sail
K   Pacific boom-lateen sail

For detailed explanations see text.

the after or lee yard-arm as the sail was adjusted (trimmed) to face the wind. But broadly speaking the limitations of these manoeuvres were soon reached, and sailors of all seas and cultures sought perennially for some escape from the essentially transverse character of the square-sail. Only by devising arrangements which would permit their canvas to be mounted more in line with the length of the vessel, i.e. fore-and-aft, could they hope to take advantage of beam or contrary winds.

All these other types of sail, which constituted successive approximations to the idea fore-and-aft rig, differ fundamentally from the square-sail in being placed with the mast closer to one edge, so that the surface area differs on the two sides. Swivelling thus round the mast as an axis, they receive the wind, as opportunity serves, now on one face, now on the other. The role of the 'sheets' or bolt ropes which are attached to the outer edge of the sail (its lee-edge or leech) to hold it in and adjust its set, becomes even more important than before. One of the most primitive forms of fore-and-aft sail is the Indonesian canted (tilted) sail, still rectangular, and shown in *B*.

The next stage may be found in the 'lateen' sail, so characteristic of the Arabic civilisation. It exists in two forms, *C* which retains a luff (i.e. a short fore or inner edge of the sail opposite the leech), and *C'*, which is purely triangular, the head of the sail joining its foot. Mediterranean and Indian Ocean lateens never have a boom, but various South Asian and Pacific peoples use sails of a triangular shape bent to booms as well as yards (*K*). These 'Pacific boom-lateens' of Indonesia, Micronesia, Fiji, etc. are believed to derive from a kind of sprit-sail (*J*) termed 'Oceanic' because characteristic of Polynesia. This in turn originated, it would seem, from a still more ancient form of rig (*H*) in which the sail, though approximately square, is held aloft by two sprits equivalent to a two-legged mast, and can thus be set fore-and-aft. This is the Indian Ocean 'bifid-mast sprit-sail' or 'proto-Oceanic sprit-sail', and it occupies a central position in the evolution of sailing craft, as we shall presently see (p. 197). It is certainly related to the strange Melanesian 'double-mast sprit-sail' (*I*), where the sail is held up by both edges. These South Asian and Pacific rigs were formerly neglected by those who wrote on the development of the sail, but it is now clear that no general scheme can omit them.

More advanced, and suitable for larger vessels, is the lug-sail or ear-shaped sail ('voile aurique'), beloved of Chinese shipmen. This is a development of both the antique square-sail and the canted sail, for it could retain the almost horizontal boom of the former but cants the yard as in the latter, while forward of the mast its luff area may be quite small (*D, D'*). The true sprit-sail (*E*) is one in which the upper angle of the leech is held out by a spar rising from near the foot of the mast; and finally the gaff-sail (*F*), or yacht-sail with which we are so familiar, is sustained by a half-sprit or gaff near the top of the mast, and a boom below like that of the lug-sail. Both of these types are fore-and-aft sails in the fullest sense, since they pivot on the mast and leave no structures forward of it. So also is the 'leg-of-mutton' sail (*G*), a triangular piece of canvas of obscure origins, secured to mast and boom alone. All else is elaboration. Jib-sails, set on a bowsprit or jib boom, are developments of stay-sails, which took advantage of the stays (the large ropes supporting the masts). Top-sails are smaller sails hoisted above the regular sails.

It will be worth while to consider for a moment the effects of wind on a fore-and-aft sail. One might easily think that the only part of the wind which strikes the windward, hollow, side of the sail has any effect; but in truth the wind which flows round the leeward, convex, side also contributes its share to the driving force. For what matters is that there should be a difference of

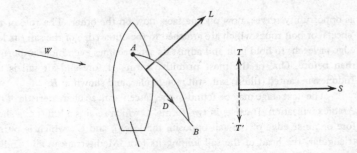

Fig. 212. Diagram to explain the action of the wind on a vessel under sail. Explanation in the text; the case considered is that of an almost beam wind and a fore-and-aft sail.

pressure between corresponding points on the two surfaces of a sail, so that a total force results. This difference of pressure is a natural consequence of the way in which wind flows past a curved obstacle such as a sail; but the curvature of the sail is less important than the fact that the wind is blowing at an angle to the boom. Indeed it is not yet generally agreed whether it is better in practice for a fore-and-aft sail to be very tightly bent to the boom so that the whole sail is as flat as possible, or for it to be allowed to set in a distinct curve. Many believe that it gives the best results when relatively taut; loose bellying sails lose energy by air turbulence, and perfectly flat ones would not give a differential flow effect.

The action of wind's force on a sail may be simply described (see Fig. 212). The wind *W* strikes the sail – or (for simplicity) the boom *AB* – at a certain angle. The difference in pressure between the two sides of the sail produces a lift force *L* perpendicular to *AB*, and a drag force *D* along *AB*. It is the conscious purpose of aerodynamic design, and has been the unconcious aim of shipbuilders throughout the centuries, that *L* should be much larger than *D*. When a boat is making good to windward, *L* and *D* may be separated into other components, *T* and *S*, *T* being the force which drives the boat forward on its course into the eye of the wind, and *S* the force which tends to drive the boat sideways – to make a leeway – a force which it is the function of the hull, keel and leeboards to counteract. But if the boat is steered too close into the wind, then the drag *D* will be so large and the lift *L* so small that the force *T* will exert itself backwards (as *T'*) rather than forwards, and the boat will make no headway to windward. One can now see why there was a historical tendency towards ever loftier sails, for the ratio *L/D* steadily increases as the height of the sail rises in relation to the length of the boom. One can also see why the loose-footed square-sail, even when braced very much forward, could not easily work to windward.

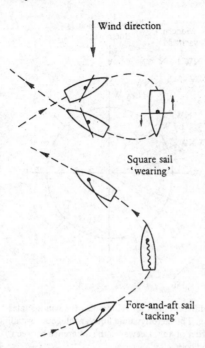

Fig. 213. Diagram to show the principles of tacking and wearing. Fore-and-aft rigs usually tack; a ship with a square-sail more often works to windward by wearing.

From the earliest days of sailing it was found that since one could not proceed directly into the wind, one must approach it by a series of passages as near the wind as possible, and as everyone knows, this zigzag movement is termed in general 'tacking'. But the movement of turn varied with the type of sail. Square-rigged ships could not often tack; they then had to 'wear about' by turning the stern to the wind (see Fig. 213), and the same is true for lateen-rigged ships, though in these the manoeuvre of tacking is occasionally done. With all the more developed types of fore-and-aft sail, however, the helm has simply to be put over bringing the bows up into the wind, and the sail 'luffs', hanging loose, until it catches the wind on the other tack. How near the courses could be to the wind is shown in Fig. 214. With all this in mind we are in a position to study the historical evidence and the Chinese contribution.

The mat-and-batten sail
The most characteristic Chinese sail is the balanced stiffened lug-sail. Fig. 215 gives a good idea of it. This shows a more or less northern type, for in

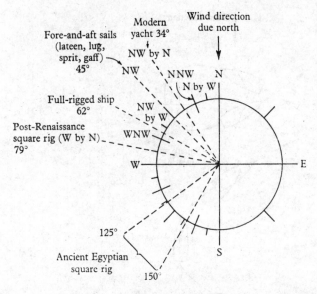

Fig. 214. The capacity of different craft for sailing near the wind; angles projected upon a wind-rose. The 'aerodynamic limit', analogous to stalling in aircraft, would be of the order of 30°. Leeway is the difference between 'sailing' and 'making a good course'.

the south the leeches of the sails are rounded. As we know from the historical material already discussed, the Chinese sail, in its most typical form, is stiffened by battens or laths of bamboo running across it, the ends of which are secured to bolt-ropes suspended from the yard so as to take the weight of what might be called the 'sail-frame', a kind of skeletal ladder. The fabric of the sail is laced to the edge of this frame and to each batten (see also Fig. 217), so that it is kept very flat. The widespread use of bamboo matting sails necessitated this sort of frame and led naturally to the balanced lug shape. The importance of tautness is, as we have seen, considerable, yet such a design, which doubtless arose because of the easy availability of a material so light and at the same time so strong, never arose in any other culture. We shall come back to this point. The battens have at least five other uses; they permit precise and stepwise shortening or 'reefing', they allow immediate furling of sail, which falls into pleats; their setting system makes it unnecessary to have cloth or canvas as strong as on other sails, and they act as steps or 'ratlines' for the crew. Above all, they are a complete protection against tearing and carrying away; a Chinese sail may have half its surface full of holes, and still draw well.

The multiple sheets or ropes attached to the sail represent a device of great interest. Each batten is connected with all the others by a system of leads

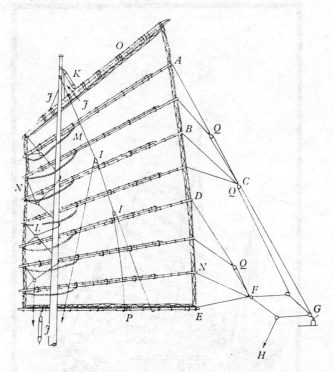

Fig. 215. Delineation of a Chinese mainsail (after G.R.G. Worcester).

ABDE	Some of the points of attachment of the multiple sheets (ropes) to the battens
ABC	Upper flexible section of sheets
DEF	Lower flexible section of sheets
CG, FG	Main sheets
G	Ring-bolt on deck
H	Feeding part of main sheets
I, I	Topping lifts and their euphroes (extended pulley blocks which include a number of holes through which separate pieces of rope may pass, not themselves on pulleys).
J, J	Mainsail halyards
K	Secondary mainsail halyards (for setting the peak or upper outer corner of the yard)
L, L	Hauling parrels
M, M	Parrels (chains)
N, N	Bolt-ropes along the free edges of the sail
O	Yard
P	Boom
Q	Euphroes of the sheets

船   條

Fig. 216. A standard transport junk in Lin-Chhing's [Lin-Qing's] *Ho Kung Chhi Chü Thu Shuo* [*Ho Gong Qi Ju Tou Shuo*] (Illustrations and Explanations of the Techniques of Water Conservancy and Civil Engineering) of 1836; the best traditional Chinese drawing which shows the system of battens, the multiple sheets with their euphroes, the halyards and the topping lifts. As usual, the stern (to the right) is higher than the bow (where the capstan and grapnel may be noted), and though a long tiller is shown the object astern is probably a sampan rather than the top of a long rudder. A transport of this kind could do coastal as well as river work.

Fig. 217. Photograph of the mainsail of a Chinese Ta-pu-thu [Da-bu-tou] freighter drawing nicely on a wet and windy day off the coat of Shantung [Shandong]. The details of the rig come clearly out. In the foreground are the multiple sheets, with two of their euphroes visible, all attached to the leech ends of the battens, where the bolt-ropes can be seen along the sail edge. Behind them to port and starboard we see three ropes of the topping lift system (*I, I* in Fig. 215), and further away to port the feed of this, which includes a twofold purchase. The halyards (*J, J* in Fig. 215) are on the other side of the mainsail, but the four main sheets of the foresail (and even the crowfoot convergence of the lower euphroe) can be seen descending to their cleats for coiling rope (belaying-cleats) on the port side of the deck. Three figures nearby give the scale. The halyards of one of the mizen masts run up on the extreme left of the picture. On the starboard side an old-fashioned lantern appears. From the Waters Collection, National Maritime Museum, Greenwich.

and loops (bights) of rope. These are gathered up by means of pulley-blocks and euphroes (extended pulley blocks so elongated as to include a number of holes through which separate loops of rope may pass, not themselves on pulleys), to form one main rope reaching to the deck, originating from a fixed point (G in Fig. 215). Thus the sail is divided into sections (e.g. *ABC, DEF*). Naturally variations in the sheeting are legion. The greater the number of battens, the greater the number of multiple sheets, the flatter the sail, and the finer the adjustment of its outer edge. The halyards, *J, K*, pass through blocks at the masthead and yard. The sail is held to the mast by a band of chain or a parrel, (*M*), for each batten, and there is a hauling parrel, *L*, arranged as shown, which assists in this and helps to raise the sail vertically when reefed. In a squall only the halyard need be touched, for the sail falls neatly into the lazy lines formed by the double upper ropes, the topping-lifts (*I, I*), the lowest batten-contained sections collapse, the ropes automatically slacken, and can easily be reset. The sail never jams. Moreover, this system avoids sending men aloft to take in the reefs, always a dangerous operation in bad weather.

Experienced Western sailors have written in glowing terms about it. H. Warrington Smyth said: 'I can speak from some experience in handling this form of sail, and I may say that once having learned the set and balance of the sails for various points of sailing, nothing can surpass the handiness of the rig'. Another expert, Captain C.P. Fitzgerald, called Chinese ships 'the handiest vessels in the world', and Admiral Paris spoke of the rig as 'one of the most ingenious of Chinese inventions'. The only criticism which he and others have made of it is that it is liable to be rather heavy.

The Chinese balance lug ranks indeed among the foremost achievements in man's use of wind power. Of the circumstances which gave rise to it we know very little, but the suggestion has been made that it arose from plaiting together successive palm branches so that the central stem or midrib provided a natural build-in batten. As we shall see in a moment our oldest Chinese text concerning fore-and-aft sailing mentions just such plaiting. The lug-sail is not the only type used by Chinese ships, however. On the Yangtze [Yangzi] the square-sail has persisted; its stiffening (accomplished by light ropes sewn in) is vertical, not horizontal, and a roller–reefing technique is used. Square sails of cloth, tall and narrow, are used on the Chhien-thang [Qiantang] River (Fig. 218). Equally interesting is the fact that the true sprit-sail exists also in China (Fig. 219). It differs from its European counterparts in being stiffened horizontally, though without battens, and since the unusual system of multiple ropes applies, vangs or ropes for holding the peak of the sail in the desired position are unnecessary. Nor do the Chinese have brails, ropes from the foot of a sail through blocks on the mast or yard so that it can be gathered for furling; this is because the sail is set or taken in by hoisting or lowering the halyards or tending the sprit. The existence of the true sprit-sail in China is

Fig. 218. Photograph of a timber freighter of the Chhien-thang [Qiantang] River, often used for transporting flood-protection faggots of brushwood etc.; one of the relatively few Chinese craft which make use of square-sails. Note also the sprit-sail on the small foremast. Photograph from R.F. Fitch.

certainly of interest, though evidence is still lacking whereby its historical significance can be assessed. It will be understood that this sail is one of those which embody the fullest development of fore-and-aft sailing, since nothing is forward of the mast.

Outside the Chinese culture-area the mat-and-batten sail as such never widely spread. It was of course used from fairly early times in Japan, where copies of twelfth-century pictures show it particularly clearly. It spread, too,

Fig. 219. A Chhangsha [Changsha] sprit-sail sampan on the Hsiang [Xiang] River photographed by Joseph Needham in 1964. Even so small a sail as this has multiple sheets.

to the Maldive islands, where it is habitually used. The Portuguese of the sixteenth century appreciated it, but there was some obstacle to its diffusion to Europe and other regions, probably the lack of bamboo or other suitable material for the battens. In 1829, however, at least one British steamship working out of Calcutta, the 'Forbes', was rigged with large Chinese batten lug-sails.

Chinese sails in history

What information can we now assemble concerning the history of Chinese sails? Among the ancient characters on the oracle-bones, *fan* (凡), later meaning 'all, every', and used as an initial particle, 'generally speaking', signified in its original form, a sail. It is interesting that this graph seems clearly to depict the 'double-mast sprit-sail' now known only in Melanesia

Fig. 220. Sketch of a Melanesian double-mast sprit-sail made at Port Moresby, Papua (after R. le B. Bowen).

(Fig. 211, *I*, and Fig. 220). Probably this was one of the contributions of the south-eastern or oceanic component of ancient Chinese culture, during the second half of the second millennium BC. It seems a less likely ancestor of the Chinese lug-sails than the canted Indonesian square-sail, but the Chinese sprit-sail could be its true descendant. The fact is that we have no clear evidence on the origin of the Chinese balance lug, but one cannot help recognising that it was the ideal answer to the problem of making headway against the monsoon winds which blow so regularly up and down the coasts of China, using a material so abundant as bamboo matting for the sail's surface.

The earliest Han mentions seem to be generally to matting, as in the expression 'blown along by mat(-sails) a thousand *li*', while the term *fan*, later usual, does not often occur with *pu* [*bu*] as *pu fan* [*bu fan*] (cloth sails) until the middle of the Later Han. Archaic ways of writing the word brought out the connection with the wind. The *Shih Ming* [*Shi Ming*] dictionary, compiled by Liu Hsi [Liu Xi] about AD 100, says that a sail is 'like a curtain, held up to the wind, so that the boat goes lightly and swiftly'. It is at any rate clear that in the last decade of the fourth century AD cloth sails were reserved for the boats of officials, indeed 'cloth sails' became a stock phrase among later poets to indicate an atmosphere of luxury, pomp, or buoyancy. Presumably mat-sails were regarded as too rough for the official junks – irrespective of the performance of the two types. Nothing here throws much light on the shape of sail or the kind of tackle used.

A third-century AD text of capital importance does so, however. It occurs in the *Nan Chou I Wu Chih* [*Nan Zhou Yi Wu Zhi*] (Strange Things of the South), written by Wan Chen [Wan Zhen], and runs as follows:

> The people beyond the barriers, according to the sizes of their ships, sometimes rig (as many as) four sails, which they carry in a row from bow to stern. From the leaves of the *lou-thou* [*lou-tou*] tree (not

easily to be identified now), which have the shape of *yung* (possibly a lattice window), and are more than one *chang* [*zhang*] (about 2.25 metres) long, they weave the sails.

The four sails do not face directly forwards, but are set obliquely, and so arranged that they can all be fixed in the same direction, to receive the wind and to spill it. Those (sails which are) behind (the most windward one) (receiving the) pressure (of the wind), throw it from one to the other, so that they all profit from its force. If it is violent, they (the sailors) diminish or augment (the surface of the sails) according to the conditions. This oblique (rig) which permits (the sails) to receive from one another the breath of the wind, obviates the anxiety attendant upon having high masts. Thus (these ships) sail without avoiding strong winds and dashing waves, by the aid of which they can make great speed.

This is indeed a striking passage. It established without any doubt that in the third century southerners, whether Cantonese or Annamese, were using four-masted ships with matting sails in a fore-and-aft rig of some kind. The writer was perhaps a little confused about the purpose of the rig, but his emphasis on the obliqueness of the sails necessitates a cut whereby they would not have got in each other's way. Additional testimony for ships with multiple masts may be derived from another third-century book now lost, passages from which are preserved in encyclopaedias.

Evidence concerning sails from subsequent centuries has been noted already in passing. Multiple masts occur again, of course, in Marco Polo and Ibn Battutah, as also in the derivative world-maps. The statement of Chu Yü [Zhu Yu] for 1090 (p. 112) that Chinese ships could use wind from almost any quarter is highly significant.

The *Kao-Li Thu Ching* [*Gao-Li Tu Jing*] (Illustrated Record of an Embassy to Korea) of 1124 is quite informative on sails and sailing. Speaking of the 'retainer ships' which accompanied the larger ships of the diplomatic envoy and carried his staff, Hsü Ching [Xu Jing] says:

When the wind blows favourably, they hoist the cloth sails (for running fast) made of 50 strips. But when the wind blows from the side they use the advantageous mat sails, set to the left or to the right like wings according to the direction of the wind. At the top of the main mast they add a small topsail, made of ten strips of cloth. This is called the 'wild fox sail', and is used in light airs, when there is almost no wind. Of all the eight quarters whence the wind may blow, there is only one, the dead ahead quarter, which cannot be used to make the ship sail. The sailors also attach some bird feathers to the top of an upright pole to act as a weathercock; this is called the 'Five Ounces'.

To get a favourable wind is not easy, so that the great cloth sails are not as useful as the mat sails, which when skilfully employed, will carry men wheresoever they may wish to go.

This important passage clearly shows that what we should now term the aerodynamic properties of taut mat-and-batten sails were appreciated in practice by a scholar early in the twelfth century, and that they were used for beating to windward, while sails of cloth, or silk (on the ambassador's ship), were hoisted additionally when running before the wind. This combination we have come across already in Fig. 182 and its description (p. 74). The use of topsails (also in Fig. 182) is interesting at this time.

Our series of texts may be completed by two passages from the sixteenth century, the first from the historian of the Portuguese conquest of the Indian Seas, de Castanheda. He spoke of the Chinese junks which had come to Malacca during the previous hundred years and earlier, bringing gold, silver, rhubarb, all kinds of silks, satins and damasks, porcelains, gilded boxes and fine furniture. They took back with them pepper, Indian cottons, saffron, coral, cinnabar and mercury, drugs and cereals.

They have but one mast, and a sail of 'Bengal matting' (made from small reeds) which turns around the mast as if pivoted on a spindle. For this reason the junks never wear around as our ships do. When it is desired to reef (the sails), it is not necessary to fold them, for they fall all in a single piece. Hence the junks sail well; . . . .

Here there can be no doubt that lug-sails were meant, as in the later description (1637) in the *Thien Kung Khai Wu* [*Tian Gong Kai Wu*] (Exploitation of the Works of Nature) of Sung Ying-Hsing [Song Ying-Xing], which runs as follows:

The sail is made by weaving together thin and narrow strips of the outer parts of the stems of bamboo, and (this matting is) divided into sections grasped by (parallel) bamboo battens. Thus the sail folds in tiers, ready to be (bent to yard and boom and) hoisted. A large mainsail in a grainship needs ten men to hoist it, but for the foresail two suffice. In order to get ready to sail, the halyards (have to be passed through) a pulley-block with a 2.5 centimetre gauge pulley fixed to the top of the mast, and brought down to the halyard winches (on deck) amidships. The adjustment of the height of the sail (according to the wind force, etc.) is thus like the variation of the three sides of a triangle. For equal sections, the upper part of the sail is thrice as effective as the lower. What matters is the adjustment to the conditions. When the wind is favourable the sail is hoisted to its full height and the boat moves at a good speed like a racing horse, but if the wind freshens the sail is reefed (coming down by its own weight) in due order (section by section one after another). [In a squall, the sail may have to be pulled down by long hooks. (Commentary by Sung Ying-Hsing).] In a gale only one or two sections of the sail are hoisted.

Lastly there follows a passage on tacking. Sung Ying-Hsing explains it clearly enough, having evidently watched the process, though he seems to have confused it a little with the handling of a ship in a current.

A beam wind is called a tacking wind. When a boat is sailing with the current, the sail is hoisted and the vessel wanders about (i.e. takes a zigzag course). Sometimes a difference of a couple of centimetres in setting the sail when tacking to the east will make all the difference between a safe passage and a setback of several tens of metres. Before she reaches the bank, the rudder is turned (hard over), and the sail is reset. The vessel is now tacking to the west, making use both of the strength of the current and the force of the wind.

The place of Chinese sails in world nautical development
How do these Chinese inventions compare with parallel progress made elsewhere in solving the problem of windward sailing? We shall try to make our answer as brief as possible, with the aid of the chart in Table 46.

There is general agreement that the oldest sailing-ships, those of ancient Egypt, were invariably rigged with square-sails. This can be seen on what is perhaps the oldest known record of sail, a painting on a piece of 1st dynasty pottery, dating from about 3000 BC. Many accept that since the prevailing wind in the Nile valley is from north to south, boats could travel downstream relying on the current, while on the return journey they could run before the wind all the way. Hence the problem of windward sailing did not arise for a long time. From this focus the square-sail radiated in all directions – to the Mediterranean, to the north, and to the whole of Asia.

It can hardly be doubted that the earliest development from the square-sail was the canted square-sail, which consists simply in so adjusting the yard arm and foot as to get the greatest possible area of the sail to one side of the mast. This is the typically Indonesian arrangement, and though evidence for this dates from the eighth century AD, Dr Needham is inclined to think that the invention is very much older there. It occurs also in the celebrated frameless boats of the Middle Nile, which may not be original.

The most prominent Western fore-and-aft sail was the lateen, so characteristic of the Islamic culture-area. The sail exists in two forms (Fig. 211), the purely triangular (C'), and the 'quasi-lug' type (C). The first of these is found only in the Mediterranean, the second throughout the Indian Ocean. Much

Table 46. *Chart showing the distribution and possible genetic derivation of types of sails*.

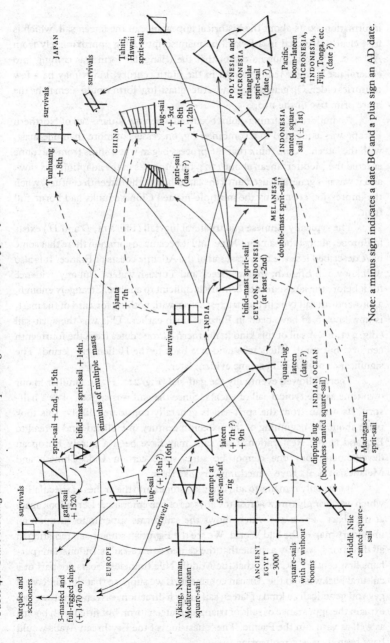

Note: a minus sign indicates a date BC and a plus sign an AD date.

information exists about the historical appearance of the lateen sail, which is first clearly depicted in a Byzantine manuscript dating approximately from AD 880. There is thus no doubt that the triangular form was coming into general use in the Mediterranean in the ninth century, but it may be a few centuries older. On general grounds, the 'quasi-lug' form would seem to be the more primitive of the two.

An important turning-point occurred when the square-sail of Northern Europe was associated in a combined rig, on vessels of more than one mast, with the lateen. Some think that the process began when ships from Bayonne entered the Mediterranean about 1304, but progress in the adaption was slow, and it was not generally used until the latter half of the fifteenth century, when the impression created by the multiple-masted Chinese junks had borne full fruit.

The typically Chinese quadrilateral lug-sail (Fig. 211, *D* and *D'*) exists in Europe, almost the same in shape, and it became so common there that some have described it as the universal sail of the Atlantic coasts of France. It is also well known in English, Italian, Greek and Turkish waters. But any evidence for it before the late sixteenth century is difficult to adduce. Strangely enough, a fore-and-aft sail type in the strictest sense, with nothing forward of the mast, is now known to have been in Europe much earlier. This was the sprit-sail (Fig. 211, *E*). A sail of this kind in northern Europe dates from the fourteenth century, but tomb reliefs take evidence back to the Hellenistic period. The significance of this will become evident later.

There followed eventually the gaff-sail (Fig. 211, *F*), so familiar in our own time as the typical sail of yachts. Since the gaff was spoken of as a half-sprit, its origin from the sprit-sail is generally considered likely; this took place about the beginning of the sixteenth century in Holland, and spread to England at the Restoration. Gaff sails may have been a purely European development, but one cannot be sure, since certain Indo-Chinese and Melanesian boats have something very like them.

It now only remains to add a few keystones into the arches of speculation which we daringly throw across the abysses of our ignorance. Let us look again at the chart (Table 46) which shows the sail forms superimposed upon an imaginary map of the Old World. We see the Egyptian square-sail radiating in all directions, we may assume that the canted square-sail is Indonesian (perhaps first century BC) and that the Middle Nile frameless boats are part of a cultural backwash to the African continent. If we suppose that the canted sail gave off genetically a kind of lateen sail in each direction, east and west, we can explain the appearance of sails of triangular lateen form, but fitted with booms as well as yards, in the Pacific. The 'quasi-lug' of the Erythrean Arabs would have been its western offspring.

There seems really little historical difficulty in believing that all lateens

Fig. 221. A fishing-junk from Tolo Harbour (the great eastern inlet in the Hongkong New Territories) using a canted square-sail as a kind of spinnaker in addition to the two usual lug-sails. From the Waters Collection, National Maritime Museum, Greenwich.

originated from the canted square-sail. We must, then, presumably suppose that the Chinese balanced lug-sail was another development from the canted rig. This process was apparently occurring at least as early as the third century AD, as we have seen (p. 193). Here the mechanics of the evolution are rather easier to understand. Besides, several traces of it have survived. For example, some of the junks of the Chhien-thang [Qian-tang] River carry square-sails, but when wishing to sail to windward, hoist them canted. Reefing by rolling the sail, a feature characteristic of the canted sail both in Indonesia and on the Middle Nile, has persisted on some of the square-sailed junks of the Upper Yangtze [Yangzi]. And certain Chinese boats carry a vestigial canted square-sail as a kind of spinnaker additional to their lug (Fig. 221). Indeed, if one cannot spend much time rambling in the South Seas oneself, one has only to look at some good photographs of the craft of Indonesia and Malaya to see the Indonesian canted rig turning almost visibly into the lug (Fig. 222).

A major problem is presented by the European lugger, with its one to three masts and its lugsails. If it really originated indigenously as late as the end of the sixteenth century, its spread around all the coasts of western Europe was astonishingly rapid. The suggestion is at least worth entertaining that it came directly from the Chinese junks, losing its batten and multiple sheets on the way. A suspicious circumstance lies in the fact that the Adriatic, Marco Polo's home waters, seems to be the geographical centre of the distribution of the rig in Europe. The *trabaccolo* and the *braggozzi*, both luggers, still predominate in Venetian ports like Chioggia, carrying two standing lugs and a jib. Still more remarkable, they have other Chinese features; they are very flat-bottomed, and they have an enormous rudder which descends some distance below the keel, and which can be hoisted up when shallow water is approached. Further hints may be gained by the fact that the only instance in Europe of multiple sheeting occurs in Turkish luggers, and there not as sheets at all, but as bowlines (ropes leading forward from the windward edge of sails, to tauten them), which with batten-sails in China had never been necessary. And on certain Turkish boats, a pair of side-wings or cheeks behind the stern, just as on the sampan, could be seen. Such indications permit us, therefore, to suggest that evidence may later be forthcoming that at some time after the return of Marco Polo to his native city, the Chinese lugsail was copied in Europe.

We return now to the first sprit-sails in Europe. Those who have discussed the matter have failed to notice that rigs of the same kind occur in China (Fig. 219). That they occur widely there prevents one from lightly assuming them to have been a transmission from Europe in the seventeenth century. The assumption often made that in Europe they must have derived from the Mediterranean lateen is also unattractive, for on general evolutionary principles the reduction of a square-sail to a triangular one would not be readily reversible. More interesting is the possibility that all sprit-sails derive from the twin-mast sprit rigs of the Indian Ocean which carry a square-sail. Many examples are known – from Ceylon, from Madagascar, from the Pacific also. In this case, the Chinese sprit-sails would be one derivative, while the European would be another.

In this connection, the discovery of the earliest depiction of a sprit-sail in Europe, is of exceptional interest. It appears on the seal of Kiel (1365), and seems to be an elongated square-sail carried on a twin (bifid) spar, though a normal mast is there as well (Fig. 223). There is apparently no yard or boom, just as in the Indian Ocean form. Furthermore, it has been found that in Tahiti and Hawaii also a conversion of the twin-mast square-sail to the sprit-sail had

Fig. 222. A small boat of the Kuala Trengganu fishing fleet (north-eastern Malaya) running for port before the afternoon wind from the sea. The foot of the canted square-sail is here placed so far to starboard, or at least amidships, that it almost approximates to a lug. Photograph Hawkins and Gibson-Hill.

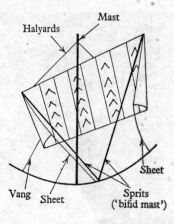

Fig. 223. The sail on the seal of the city of Kiel (1365). For explanation see text.

occurred, perhaps independently. Again there may be significance in the fact that one of the chief foci of the sprit-sail in Europe is Turkish.

It may be that Indian influence on Mediterranean sails was exerted a thousand years earlier than indicated by the ship in the seal. Four Hellenistic tomb reliefs of about the second century AD do seem to represent sprit-sails of some kind or other, though two of them resemble the seal's arrangement more closely than the true sprit-sail of later times. The lack of any evidence for the long intervening period remains however somewhat mysterious, and the suggestion that there were two successive introductions cannot be excluded. That adequate Roman–Indian contacts for the exchange of techniques existed cannot be doubted. It is now also clear that the Indian Ocean twin-mast could not possibly have been derived from Europe, for its luff is loose, and by the time that the Europeans came again to Indian waters in the sixteenth century, their sprit-sails had long been laced to the mast. Indeed, the Indian Ocean is coming to occupy a central position in the evolution of the sprit-sail; its early twin-mast sprit-sail could have been an ancestor of the triangular sprit-sails and boom-lateens of Oceania, as well as of the Melanesian 'double-mast' sprit-sail which linguistics links with the earliest sails of China.

One last sail has so far not been mentioned: the triangular peakless fore-and-aft sail (Fig. 211, *G*). This has a modern air because it is used on racing yachts, where it is called the Bermuda rig or 'leg-of-mutton' sail. Its wide application in modern times has come about because of the realisation that most of the work of a sail of optimum rigidity is done by the leading edge, and its physical action is similar to that of a bird's wing. But sails of this triangular shape are very much older, for it is found in the Chinese culture-area, notably on some Indo-Chinese craft. Here the yard is lashed to the mast and stands upright and the boom converges to it with or without a very small luff. It might be regarded as an 'Arab' lateen sail shifted horizontally leech-wards so far that its yard could almost become a vertical continuation of the mast; and since indeed it was almost certainly derived from standing lug-sails of more normal shape, this is what it really is. Of course the presence of the boom betrays its origin from the canted Indonesian square-sail. In short, the leg-of-mutton sails of Europe and Indo-China both ultimately derived from the canted Indonesian square-sail through the convergence of the lug's yard and boom at the luff edge.

It would be hard to summarise all the complex facts which we have touched upon here. Yet is seems sure enough that after the first development of the canted square-sails, wherever and whenever that may have taken place, the earliest fore-and-aft rigs appeared in the Indian Ocean about the time of King Aśoka (third century BC), spreading in different forms both to the Mediterranean area and to the South China Seas by the second century AD. Seemingly in the West the sprit-sail was forgotten, but in East Asia the tall balanced lug-sail went from strength to strength. Next following were the lateen sails of Arabic culture from about the seventh century AD onwards. European sailors of medieval times were slow to adopt these more advanced techniques, the fifteenth-century Portuguese pioneered nobly in windward sailing. Whether or not the European lug-sail was derived directly from China or South East Asia remains to be proved, though it seems exceedingly likely. At all events the credit for the oldest decisive inventions which permitted more and more efficient sailing near the wind must go to the peoples of Indian, Indonesian and especially Sino-Thai culture.

Leeboards and centre-boards.

In all fore-and-aft sailing there is likely to be a marked tendency for the ship to drift to leeward (see p. 184). The tendency can only be counteracted by making the hull of such a shape as to minimise this drift. Here the western keeled construction had great potential advantage over the Chinese flat-bottomed box build. But other means of opposing leeward loss may take the form of movable flat boards, either attached to the side of the vessel or let down through a water-tight channel or 'trunk' along the midship line – they are known as leeboards and centre-boards.

Many types of Chinese river craft are equipped with these devices. Such leeboards tend to be about one-sixth the length of the ship, and occur in various fan-like shapes. The Cantonese sea-going trawlers have conspicuous centre-boards. Japanese drawings of the eighteenth century show leeboards on ships from Nanking [Nanjing]. Indeed there is good reason for thinking that the Chinese culture-area is their original home, for as Hornell recognised,

the Taiwanese bamboo sailing-raft, perhaps the most ancient existing prototype of all Chinese vessels, has both leeboards and centre-boards. These are
essential for its work, (it can actually beat to windward with its lug-sail), and
though now assisted by steering-oars they are capable of carrying all the
burden of the helm, as they still do in its Peruvian counterpart. Then as
Chinese shipbuilding developed it would have been most natural to continue
the use of leeboards when the flat-bottomed junk construction came to be
powered by efficient fore-and-aft rigs.

    While medieval Chinese references to them are rare, a rather clear one
can be seen in the description in the military treatise *Thai Pai Yin Ching* [*Tai
Bai Yin Jing*] (Manual of the White (and Gloomy) Planet (of War; Venus)) of
AD 759. A passage which we shall quote more fully later (p. 259) says that
there are attached to the hull on both sides 'floating boards' shaped like the
wings of a bird (a hawk or grebe) after which this class of vessel was named.
These attachments 'help the ships, so that even when the wind and wave arise
in fury, they are neither (driven) sideways nor overturn'. The writer Li
Chhüan [Li Chuan], may not himself perhaps have been perfectly clear about
the function of these devices, but what he says can hardly refer to anything but
leeboards. There is no evidence for their existence in Europe before about
1570 at the earliest. This has led to the suggestion that the device was brought
to Europe from Chinese waters. In favour of this view, perhaps, would be the
fact that then they appeared first upon Portuguese and Dutch ships.

    For writers of the Ming leeboards were a commonplace. Speaking of
river and canal ships Sung Ying-Hsing [Song Ying-Xing] says:

> If the ship is rather long (in relation to the depth of the rudder with
> which she is fitted), then in a strong cross-wind the power of the
> rudder will be insufficient (to prevent leeward drift). Then broad
> boards are lowered into the water as quickly as possible, and this
> counteracts the influence (of the wind).

And of sea-going ships:

> ... Amidships there are large horizontal beams set thwartwise and
> projecting outboard a metre or so; these are for letting down the
> 'waist-rudders' (centre-boards). All sorts of ships have them. These
> waist-rudders do not resemble the proper stern rudders in form, but
> are fashioned out of broad boards into the shape of knives, and when
> lowered into the water they do not turn, but help to keep the boat
> steady. At the top there are handles fitted on to the cross-beams.
> When the ship sails into shallow waters, these waist-rudders are
> raised, just as the rudders themselves are, hence the name.

Curiously relevant here is the quotation mentioned above (p. 195) in another connection from de Castanheda. Referring to the period 1528 to 1538 he wrote, just before the section quoted, of junks having two rudders, one at the stern and another at the bow. What perhaps he or his informants had actually seen were certain southern vessels, such as the *tang-way* [*dang-way*] of Kuang-tung [Guang-Dong], and the *ghe-nang* of Annam, which have a movable board sliding up and down at the bows. This bow centre-board does give, from a distance, the impression of a rudder, and of course the lee- or centre-board, in any form, is practically unknown in traditional Mediterranean craft.

An alternative method of accomplishing the same end consists in having the rudder relatively large, and lowering it well below the bottom of a flat-bottomed craft. This was certainly one of the reasons for the widespread Chinese medieval use (which still continues) of rudders deeply slung, and capable of being raised when navigating shallow waters.

OARS

Rowing and the handled oar

Since paddles and oars used by rowers go back to prehistoric times, it is quite natural to find that the Chinese language contains a number of terms for them. The exact meaning of these is subject to local variation, and it is hard to identify the significance that they had two thousand years ago. Perhaps the central word is *chi* [*zhi*] (楫) or *chieh* [*jie*] (檝); this goes far back into the first millennium BC being found in the *Shih Ching* [*Shi Jing*] (Book of Odes) and the *Shu Ching* [*Shu Jing*] (Historic Classic). More clearly defined is the steering-oar or stern-sweep, *shao* (艄), a word which is differentiated from that for the stern of a boat, *shao* (梢). The term *jao* [*rao*] (or *nao*) (橈, 蕘) originally meant a curved piece of wood, and may therefore have been an early name of the curved 'propeller' or angle-oar which we shall discuss shortly. The name for this has long been *lu*, a word which can be written in a number of different ways. Lastly, the punt or quant pole, *kao* [*gao*] (篙) has always been clearly distinguished.

Since in Chinese vessels the deckhouses were always placed towards the stern, while the rowers took their places on the deck forwards, the word *lu* (written in yet another slightly different manner) came to mean the bows. This is evident from Han literature, a study of which would throw much light on nautical practices. It occurs in a text from 106 BC, and in the accounts of the fighting that accompanied the restoration of the Han dynasty, about AD 33, for the biography of the general Tshen Phêng [Cen Peng] tells us that 'There were on the transport ships more than 60 000 rowers working the oars (*chao* [*zhao*]).... Several thousands of boats with *lu-jao* [*lu-rao*] rushed to the attack'. The Thang [Tang] commentator Li Hsien [Li Xian] explains the latter terms as meaning that the oars (*chieh* [*jie*]) were outside the ship while the men were inside. This seeming platitude probably implies that the rowers were invisible from outside because of built-up gunwales for protection against arrows, the oars issuing through ports (Fig. 196).

Probably in the Han time, as still today, Chinese sailors generally rowed standing and facing forwards. This was fairly common ancient Egyptian practice, and in China would have arisen naturally from the paddling of the long South East Asian canoes which gave rise to the dragon boats. Oars are sunk much more deeply in the water at each stroke than is customary in Europe, and often carry a transverse handle at the end of the upper end of the oar, convenient for this type of rowing. These T-handled oars, go back at least to the Sung [Song], as we know from paintings. Facing rowing has a wide distribution, and though not characteristic of Europe, occurs in connection with Venetian gondolas and the boats of Austrian and Hungarian lakes.

During the Sung [Song], Yuan and Ming periods quite large ships were often classified according to the number of oars they carried. Thus speaking of grain-transports called 'wind-borers', surely in-allusion to their quality in beating to windward, a description from 1275 of Hangchow [Hangzhou] towards the end of the Sung divides them into greater and smaller 'eight-oared' types, besides some with six oars only. These were almost certainly not ordinary oars but large sweeps like the 'yulohs' next to be described.

Sculling and self-feathering 'propeller'

Besides the propulsive effect produced by oars worked in the ordinary way, motion forward can also be obtained by mounting an oar approximately in the line of the main axis of the boat, most conveniently at the stern, but also elsewhere, and moving it from side to side of that axis. This technique is used only for small dinghies in the West, but the Chinese elaborated it into an extremely ingenious heavy duty system. Known as the 'yuloh' (actually a verb *yao lu*, 'to shake the oar'), it has often been described; we have already had occasion to refer to it several times. Its great effect is due to three special

Cup and pin system at the top of the post supporting the yuloh.

Fig. 224. Working a small lighter (probably at Canton) by the use of the self-feathering scull oar. The motion of this yuloh, which pivots on a fulcrum and is attached to the deck by a short length of rope, amounts to that of a reversible screw propeller. Photograph R.F. Fitch, 1927.

modifications: (*a*) the loom or handle is brought more nearly parallel to the deck by means of a curve or angle inboard of the pivot-point or fulcrum; (*b*) the fulcrum consists of a peg or thole-pin fitting into a hole in a block attached to the oar; and (*c*) the handle is attached to a fixed point on the deck by a short length of rope. The cup and pin system at the top of the yuloh constitutes an approach to the universal joint. It is obvious from a consideration of the forces involved of such a 'sculling-oar' that the effect desired is accomplished only during a certain part of the stroke, and for the remainder the oar must be feathered so as to offer no resistance to the water. The yuloh accomplishes this without any need for work with the wrist; it is simply pushed and pulled to and fro, the handle necessarily describing an arc of a circle, while at the moment required for the feathering, the rope is given a jerk (Fig. 224). The whole process is difficult to describe, but can be quickly appreciated when it is seen in action.

How old the yuloh system is cannot be said with certainty but it seems to go back as far as the Han. The word *lu* itself is Han. The *Shih Ming [Shi Ming]* (Explanation of Names) dictionary of Liu Hsi [Liu Xi], dating from about AD 100 says: '(That which is) at the side (of the boat) is called *lu*. Now *lu* is connected with *lü* (the backbone). (Thus) when the strength of (men's)

Fig. 225. 'Boats descending the Gorges out of Szechuan [Sichuan]', a painting by Li Sung [Li Song], about 1200 (Ming copy). The slung rudder and the outboard thole-pins for the yulohs are noteworthy.

spines is used, then the boat moves (forward)'. This does not exclude the self-feathering propulsion oar because it has long been the practice in China to mount two or more of them near the stern or even near the bows (see, for instance, Fig. 225).

One of the first references to the *yao-lu*, as a phrase, is in the *Fang Yen* [*Fang Yan*] (Dictionary of Local Expressions) by Yang Hsiung [Yang Xiong]. As the book may contain some later interpolations we cannot be quite sure that this is the original text of 15 BC. However, it would certainly indicate a time not subsequent to the Later Han. The passage says: 'That which sustains an oar (*cho* [*zhuo*]) is called a *chiang* [*jiang*]. *Chiang* – a small wooden peg for the *yao-lu*. . . . That which ties down the oar (*cho*) is called the *chhi* [*qi*]. *Chhi* – the rope attached to the head of the oar.' *Yao-lu* appears again in the accounts of the battles between the States of the Three Kingdoms period. When Lü Mêng

[Lü Meng], a general of Wu State, was campaigning against the state of Shu he employed at one point (AD 219) a ruse in which he dressed his best soldiers in white like merchants and mounted them on merchant ships which they sculled back and forth. Since the text dates from before the end of the same century, when many contemporary documents must still have been available, it is rather firm evidence for the dating of this technical phrase.

Of non-Chinese sources, Ibn Battutah gave an excellent description which we have already quoted (p. 117), and the yuloh much impressed the Jesuit missionary Louis Lecomte, who wrote:

> ... As for ordinary Barks, they do not row them after the European manner; but they fasten a kind of long Oar to the Poup, nearer one side of the Bark than to the other, and sometimes another like it to the Prow, that they make use of as the Fish does of its Tail, thrusting it out, and pulling it to them again, without ever lifting it above the Water. This work produces a continual rolling in the Bark; but it hath this advantage, that the Motion is never interrupted, whereas the Time and Effort that we employ to lift up our Oars is lost, and signifies nothing.

The yuloh also impressed the British Navy, for we learn from Admiralty papers that in 1742 experiments were made in which a sloop was fitted with 'a set of Chinese sculls'. Something rather similar to the yuloh figured among the nautical inventions of the third Earl of Stanhope about 1790; it was called the 'Ambi-Navigator or Vibrator'. The intention was to apply steam power to the device, but it never proved successful. Then in 1800 Edward Shorter arranged 'a two-bladed screw, at the end of a revolving shaft, set at an angle like an oar in sculling, and having a universal joint connection with a horizontal shaft on deck'. Two years later a deeply laden naval transport, worked by eight men operating a capstan, made $1\frac{1}{2}$ knots (2.8 km/hr) with this device. There is also the curious story reported by the historian J. McGregor, that 'a Chinese screw-propeller was brought to Europe and seen by a Col. Baufoy in 1780', some 30 years after the Swiss mathematician and physicist Daniel Bernoulli had first suggested the possible superiority of the screw over the paddle-wheel. It may well be that the yuloh played a certain part in the early inspiration of screw propulsion.

The human motor in East and West

Naval archaeologists have busied themselves with detailed controversy about the construction of Greek and Roman galleys, the interpretation of the technical terms for the banks of oars, and what is to be made of the apparent existence of galleys with 40 banks superimposed. The only relevance this has for us is that it prompts us to ask why such problems never arose in the Chinese culture-area. Apart from the dragon-boats which were used only for racing in folk-festivals, the many-oared galley (though obvious prototypes in the long canoes of the south-east, with their numerous paddlers, were close at hand) remained throughout recorded history absolutely foreign to Chinese civilisation.

The conclusion that this indicates a clear technical superiority of Chinese seamanship seems almost unavoidable. Good reasons have already been given for believing that South East Asian waters were the scene of the earliest really successful attempts at sailing into the eye of the wind. If the first lug-sails and sprit-sails were of the second and third century AD, they ante-dated the lateens by four or five hundred years, and gaff-sails by seven hundred or more. After all, the object of using the human motor on a mass scale was not only to overcome flat calms, but to make headway against strongly adverse winds. The greater efficiency of the rigid Chinese mat-and-batten sails must at least have reduced the number of occasions when recourse to other motive power was essential. The naval requirements were not so widely different; there is plenty of combat afloat recorded in the Chinese official histories. And when the Chinese did turn their attention to other motive power, they pro-duced, with that mechanical inventiveness only grudgingly accorded to them by other peoples in modern times, the subtler yuloh (from about the second century AD), and then the treadmill-operated paddle-wheel boat (at least from the eighth, if not before).

There is nothing here in contradiction with the fact that human motive power was used on a tremendous scale in ancient and traditional China. All sailors rowed, every seaman could handle the yuloh, but their work was not done under conditions of organised inhumanity which often prevailed in Europe. There were also the trackers, whose work along the banks of the great rivers was sometimes gruelling, but again in China they were not slaves. As we have already seen, some Chinese ships were named from the number of their 'oars' (the term includes yulohs). When these were quite large, as some of them certainly were, the gear was carried primarily for manoeuvring in landlocked waters, or when way was urgently required in a flat calm. Moreover, there were many kinds of fast official-carrying boats (like admirals' gigs) and patrol boats, which were thus powered. But put it all together and it adds up to nothing approaching the galley-slave pattern characteristic of Mediterranean Europe for some two thousand years.

Athenian galleys, which were equipped only with steering oars, were rowed by free citizens, but in the seventeenth century AD the galleys which all had axial rudders, were propelled by slaves under some of the worst conditions ever recorded in the annals of slavery. All the same, it is clear that the coming of the axial rudder, which permitted great increases in the size of sailing-ships, was one of the factors which did eventually lead to the final disappearance of the rowed galley and the merciless exploitation of the human motor. There were, of course, other factors too, probably at least as important, such as advances in shipbuilding which multiplied the number and height of the masts; and especially the use of gunpowder, since cannon could devastate an attacking fleet of galleys, and they themselves being long and narrow, could not support the relatively stable platform which gunners required. It is a remarkable fact that each of these technical developments, if traced back to its origin, can be shown, with probability or certainty, to have emanated from Asian, usually Chinese, civilisation. Some of these demonstrations have already been given; others will follow in due course.

6

Steering

INTRODUCTION

As anyone who has paddled a canoe will know, the simplest form of
steering-gear is an oar or paddle held steady at a desired angle on either after
quarter (and most conveniently on the right or starboard quarter). In this
way, the streamline flow of the water is deflected so as to impart a turning
motion to the hull. One the other hand, the most highly developed form of
steering-gear is a great vane pivoting upon the stern-post of a ship, controlled
from its bridge by means of chains to which a source of power is applied and
which thereby substitute more effectively for the leverage exerted in small
ships by the bar known as the tiller.

In the West, the terminology of successive stages in direction-control
presents little difficulty. First there were steering-oars or quarter-paddles, or
occasionally stern-sweeps centrally fixed; then came rudder-shaped paddles
permanently attached to the stern quarters; and lastly the stern-post rudder
itself, hung on a hinge known as a pintle and gudgeon. But Chinese terminology is a more difficult matter, since the thing changed while the words did
not.

It has been claimed that the weakness of the steering-oar was a cardinal
limiting factor to nautical development until the thirteenth century AD. Until
that turning-point the capacity of ships was restricted to about 50 tonnes. Lack
of manoeuvrability also kept them slow, and the fact that in heavy weather
any kind of steering-oar would inevitably take charge, interfering with the
handling of the sails, meant that ships were constrained to keep within reach of
shelter and could not venture to any extent on ocean passages.

The steering-oar, however, has always remained of value in rapid rivers
and narrow land-locked waters, hence its continued use in China today. To
respond to the rudder, a boat must be moving relative to the surrounding
water, for otherwise there is no streamline flow to be diverted. But when
descending rapids, a boat may be moving at almost the same speed as the
water. In such a case it is highly advantageous to have a long stern sweep so

that its effect depends not on the streamline flow but on reaction to water resistance, just as in the case of an ordinary oar. The lever, in such a stern sweep, is much longer on each side of the fulcrum than it is in the rudder. Imparting to the boat's stern a strong crossways movement, it can equally well be used for turning the vessel about when stationary in a lake or harbour.

Of all the wonders of China which so impressed Louis Lecomte at the end of the seventeenth century, none was greater than the seamanship of the river-junk men, and this he recounted in a vivid and eloquent passage:

> ... The knack the Chinese have to sail upon Torrents is somewhat wonderful and incredible; .... Had I not been upon these perilous Torrents my self, I should have much ado to believe, upon another's report, what I my self have seen. ...

After describing how he and another Jesuit were nearly lost on the way to Canton and the dangers of the cataracts in Fukien [Fujian] province, he describes the watertight compartments of the boats. He then continues:

> ... For to moderate the Rapidity of the Motion, in places where the Water is not too deep, six Seamen, three on each side, hold a long Spret or Pole, thrust to the bottom, wherewith they resist the Currant, yet slackening little by little, by the help of a small Rope made fast at one end of the Boat, and twin'd at the other round the Pole, that slips but very hardly, and by continual rubbing, slackens the motion of the Bark, which, without this Caution, would be driven with too much Rapidity; insomuch that when the Torrent is even, and uniform, how precipitous soever its Course be, you float with the same slowness, as one does upon the calmest Canal; but when it winds in and out, this Caution is to no purpose; then indeed they have recourse to a double Rudder, made in fashion of an Oar, of forty or fifty Foot long, one whereof is at the Prow, and the other at the Poop. In the plying of these two great Oars consists all the Skill of the Sailors, and the safety of the Bark; the reciprocal Jerks and cunning Shakes they give it, to drive it on, or to turn it right as they would have it, to fall just into the Stream of the Water, to shun one Rock, without dashing on another, to cut a Currant, or pursue the fall of Water, without running headlong with it, whirls it about a thousand different ways – It is not a Navigation, it is a Manège; for there is never a manag'd Horse that labours with more fury under the hand of a Master of an Academy, than these Boats do in the hands of these Chinese Mariners: So that when they chance to be cast away, it is not so much for want of Skill as Strength; and whereas they carry not above eight Men, if they would take fifteen, all the Violence of the Torrents would not be capable to carry them away.

Fig. 226. A boat being tracked upstream in one of the rapids in the Yangtze [Yangsi] gorges near Chungking [Chongqing]. Photograph by Potts, about 1938).

Lecomte's description, so graphic as to be almost incoherent, is not at all exaggerated, as may be seen from accounts of contemporaries and, Dr Needham remarks, from his own experiences. A picture such as that in Fig. 226 shows a boat negotiating rapids in the Yangtze [Yangzi] upstream. It illustrates very well the kind of conditions which must have sharpened men's minds to seek ever more effective vessel control, some specific needs leading to the development of the steering-oar into the greater stern-sweep, and others leading to the invention of the axial rudder.

The limitations of the steering-oar or stern-sweep became particularly severe at sea, or upon great lakes where rough water was likely to be met with. A ship of any size required a very considerable spar for this duty, and all the worse the consequences would be when it broke under the impact of heavy seas. Other disadvantages attended upon the attachment of a short but heavy paddle to the aft starboard quarter; it made an inconvenient projection liable to foul other ships or come into collision with quays, and that this was felt is shown by the fact that the Roman vessels were characteristically built with a kind of streamlined shield to protect the quarter-paddle. The chief value of the quarter-paddle lay in its balanced character, a flat blade existing on both sides of the axis, and not on one side only, but the universal adoption of the stern-post rudder in the late Middle Ages shows how the weight of advantage lay, and in China, as we shall see, the developed axial rudder retained its balanced form.

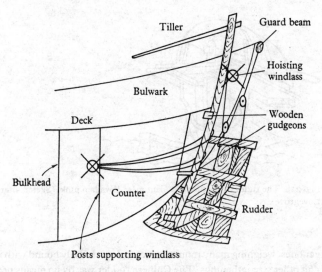

Fig. 227. The rudder and tiller of a Hangchow [Hangzhou] Bay freighter.
(After D.W. Waters).

We must now pause for a moment to consider what is known of the way in which rudders were attached to Chinese and European ships, and how this is still done in shipbuilding of the traditional kind. In Western antiquity steering-oars and quarter-paddles were slung in various forms of tackle, probably in China also, for China remains the realm of the slung sliding rudder *par excellence*. Indeed there is no evidence that Chinese rudders were ever attached by eyes or gudgeons (rings) so as to hinge with pintles (pins or bolts) on the hull. In Western ships and boats the pintle was always erect, standing parallel with the stern-post and pointing upwards if attached thereto, downwards if attached to the rudder itself. Such hooks and eyes were foreign to Chinese usage. Throughout the ages their rudders have been held to the hull primarily in wooden jaws or sockets, and, if large, suspended from above by a tackle pulling on the shoulder so that they can be raised or lowered in the water (Fig. 227). Sometimes the foot of the largest type of rudder is even connected with the fore-part of the ship by tackle which holds it in place (see Fig. 184). The gudgeon-like fittings (bearings, as it were for the main rudder post) can be open, half-open, or occasionally altogether closed by outer pieces of shaped timber (Fig. 228). Thus they have corresponded to the braces (eyes, gudgeons) of Western ships, though what turned in them was not the pintle but the rudder-post itself. The system of open jaws was what permitted the rudder to be raised and lowered.

Though cable and wood thus took the place of iron hinges, it should not be thought that iron was absent from rudders traditional in China, for in fact

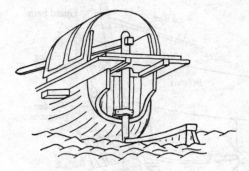

Fig. 228. The rudder and tiller of a Cantonese Kwailam junk. (Sketch after H. Lovegrove).

the larger ones, weighing many tonnes, were, and are, heavily bound with iron straps and other strengthenings. The Chinese rudder was by no means necessarily located at the aftermost point of the hull or upperworks, but sometimes considerably forward of this, its post indeed frequently descending through a rudder-trunk built into the hull. Such a construction was facilitated by the cross-piece and bulkhead anatomy so typical of the Chinese junk, and we shall see in the dénouement how closely the whole conception of the vertical steering mechanism was connected with this. The essential point is that the rudders of Chinese ships, always remarkably large in relation to the total size of the vessel, were in principle vertical, axial, and median. They were in fact 'stern-post rudders' without the stern-post. To this paradox we shall return.

To end this introduction let us quote a few words from two seventeenth century witnesses, both of whom are already familiar friends. In *The Exploitation of the Works of Nature* Sung Ying-Hsin [Song Ying-Xing] wrote:

... a rudder is constructed to divide and make a barrier to the water, so that it will not itself determine the direction of the vessel's motion. As the rudder is turned, the water turbulently presses on it, and the boat reacts to it.

The dimensions of the rudder should be such that its base is level with the bottom of the (inland transport) ship. If it is deeper, even by a couple of centimetres, a shallow may allow the hull to pass but the stern with its rudder may stick firmly in the mud.... If the rudder is shorter, even by a couple of centimetres, it will not have enough turning force to bring the bows round.

The water divided and obstructed by the rudder's strength, is echoed as far as the bows; it is as if there were underneath the hull a swift current carrying the vessel in the very direction desired. So nothing needs to be done at the bows. All this is marvellous beyond words.

The rudder is worked by a tiller attached to the top of its post, a 'door-bar' (as the sailors call it). To turn the boat to the north the tiller is turned to the south, and *vice versa* ... The rudder is made of a straight post of wood [more than 3 metres long and 1 metre in circumference for grain-ships] with the tiller at the top, and an axe-shaped blade of boards fitted into a groove cut at its lower end. This blade is firmly fastened to the post with iron nails, and the whole is fixed (with tackle) to the ship to perform its function. At the end of the stern there is a raised part (for the helmsman) which is also called the 'rudder-house'.

This is a fresh-water sailor's description, and Sung himself must often have stood beside the helmsman on the Poyang Lake or the Grand Canal; his attempt to describe streamline flow is of particular interest. Half a century later Louis Lecomte wrote likewise of sea-going junks:

Their Vessels ... are all flat-bottom'd; the Fore-castle is cut short without a Stem, and the Stern open in the middle, to the end that the Rudder, which they shut up in a Chamber, may be defended on the Sides from the Waves: this Rudder, much longer than ours, is strongly ty'd to the Stern-post by two Cables that pass under the whole length of the Vessel to the fore-part, and two other suchlike Cables hold it up, and facilitate the hoisting or lowering it, as occasion serves. The Bar [i.e. the tiller] is as long as is necessary for guiding it; the Seamen at the Helm are also assisted by Ropes fastened to the Larboard and Starboard, and roll'd upon the extremity of the Bar they hold in their Hand, which they fasten or slacken as they see occasion, to thrust or stop the Helm.

Lecomte thus aids us by showing that some mechanical assistance or advantage was applied to the tiller in large junks in his time,

FROM STEERING-OAR TO STERN-POST RUDDER IN THE WEST

The stern-post or axial rudder began to appear in Europe in the thirteenth century; before that time there was no trace of it in the West. Steering-paddles, quarter-paddles connected by a framework and bar, and the stern-sweep were known and used in ancient Egypt, while steering-oars were familiar on Greek, Roman and Hellenistic ships, sometimes single, sometimes double and also united by a bar. Evidence exists that they now began to be slung permanently in position. Byzantine culture developed nothing new, but the Viking long-ships went on to attach a pivot to the steering-oar, and thence to develop the paddle into a rudder-shape and hinge it on to the side of the boat. Norman ships continued the tradition, but up to the end of the twelfth century and beyond, the steering-oar was still illustrated by artists and sculptors. The oldest European manuscript illustration of a stern-post rudder, with tiller, dates from 1242. However, a notable iron-bound rudder on a ship is depicted on the seal of Ipswich which came into use close to 1200, and ships carved on fonts in' Belgium and England show them; these last were the work of a school of artisans from Tournai, and are datable at about 1180. Subsequent developments related rather to the control devices than to the rudder itself.

In sum, therefore, we may take 1180 as the date-line for the first introduction of the stern-post rudder in Europe. And it is quite noteworthy that this is within a very few years, perhaps less than a decade, of the first mention of the mariner's compass in the West. This fact alone might arouse one's suspicion that the stern-post rudder was not an independent development either, but made its appearance as the result of long travel from somewhere else.

CHINA AND THE AXIAL RUDDER

The problem of the history of the rudder in China presents us with a classical case of the difficulty which arises when there is reason to believe that one single word has done duty through the centuries for two or more devices technologically quite distinct. The word *tho* [*tuo*] (or *to* [*duo*]) (柁, 舵, 柂, 杝) certainly meant 'steering-oar' or 'steering-paddle' in the third century BC; equally certainly it meant the axial 'stern-post' rudder in the thirteenth century AD. Since, as we have just seen, the first appearance of the latter in Europe antedates 1200 by very little, any investigation of a possible Chinese contribution cannot rely upon names and words alone. It is necessary to see what everyone who used the word actually said about the thing.

(i) Textual evidence

The simplest procedure is to group the relevant texts into separate classes. One of the earliest must be that in the *Huai Nan Tzu* [*Huai Nan Zi*] (The Book of (the Prince of) Huai-Nan) of about 120 BC, and here the archaic word *to* [*duo*] (杕) is employed. But there are others from the first century BC, onwards, and one, from 15 BC, the *Fang Yen* [*Fang Yan*] (Dictionary of Local Expressions) says that 'The stern of a boat is called chu [zhu]. The *chu* [*zhu*] (軸) controls the water.' Of this Kuo Pho [Gou Po], a commentator from the fourth century, says 'Nowadays, Chiang-tung [Jiang-dong] people pronounce it like *chu* (軸) (axle)'.

Though some passages give no technical help at all, this remark by Kuo

Pho [Guo Po] about the local pronunciation of a word where both characters share the same phonetic, and differ only in root, makes it seem that our scent grows warm. For here is something 'turning' like an axle at the stern. But evidence is still insufficient to distinguish between the steering-oar or stern-sweep pivoted on a short boom or rowlock and the rudder held in bearings like those of a rotating shaft or axle.

We can, perhaps, make headway a little further by considering the verbs used. The word *chuan* [*zhuan*] (轉), for instance, has the nuance of something swinging round an axis rather than of something pivoting on a single point. And we find it in a passage of the third century AD in a story about Sun Chhüan [Sun Quan], later emperor of the Wu State, who forces a change of course: '... (the helmsman) immediately turned the helm (chuan tho [zhuan tuo]; whatever it was), and sailed into Fan-khou [Fankou]'. This verb came naturally to the thirteenth-century poet Yü Po-Sêng [Yu Bo-Seng], who used exactly the same impression for what was by then (1297) undoubtedly an axial rudder. But a more important piece of evidence occurs in a poem by the great radical minister Wang An-Shih [Wang An-Shi] (1021 to 1086) where when writing 'turning their rudders from east to west, the ten thousand ships returned', he uses the word *lieh* [*lie*] (捩). The axial force of this word can be appreciated from the fact that in the seventeenth century it was used for a pulley-block, and in multifarious combinations of engineering technical terms, including pivots, much earlier than that.

Next comes the consideration of shape and length. In a lost book of the fifth century AD there is a text which says '... A *tho* [*tuo*] was once seen floating downstream (from a lake).... Men sent ... found the remains of a flat-bottomed skiff by the lake, ...' This implies that it was then quite possible to distinguish a rudder from an oar by its shape. This would not have been the case if the steering-oar had still been the only device in use, for it would not readily have been distinguished from any other oar, and one of the ordinary words for oar would have been employed. So by about 450 at any rate there was a clear difference in shape. More important, perhaps, is a passage in an obscure book we have come across in connection with the history of the magnetic compass, *Master Kuan's [Guan's] Geomantic Instructor* which cannot be earlier than the seventh century but is not later than the end of the ninth. In speaking of the proper depth for tombs the writer says that 'If the *tho* goes deeper than it ought to, then the end of the boat will not carry its cargo (because it will go aground or strike a rock)'. This gives practical certainty, for the typical Chinese rudder has always been slung adjustably so that it can hang down well below the level of the ship's bottom, and aid in preventing leeward drift.

In the next century about 940, Than Chhiao [Tan Qiao] in the *Hua Shu* (Book of the Transformations (in Nature)) wrote about the length of a rudder; his value of 2.5 metres is clearly much too short for a steering-oar or stern-sweep (often over 15 metres on comparatively small river boats), but agrees well with an axial rudder. The most decisive passage, however, occurs in the *Hsüan-Ho Fêng Shih Kao-Li Thu Ching* [*Xuan-He Feng Shi Gao-Li Tu Jing*] (Illustrated Record of an Embassy to Korea in the Xuan-He reign period) written by Hsü Ching [Xu Jing] about the mission of 1124. It is full of references to rudders, often to mishaps when they broke, or to changing them, but the main sentences are these:

> At the stern there is the rudder (*chêng tho* [*zheng tuo*]) (正柂) of which there are two kinds, the larger and the smaller. According to the differences in the depth of the water, the larger is exchanged for the smaller, or vice versa. Abaft the deck-house two oars are stuck down into the water from above, and these are called 'Third-Assistant Rudders'. They are only used when the ship begins to sail in the Ocean.

These passages leave no doubt in the mind that by the beginning of the twelfth century on Chinese ships several sizes of axial rudders were carried and used under different conditions, while at the same time steering-oars may have been retained for special purposes. This was sometimes done long afterwards in European vessels.

Korean craft of the early twelfth century also had axial rudders, for Hsü Ching [Xu Jing], describing the coastguard patrol boats which came out to welcome the admiral's flotilla, says that they were one-masted, with no deck-house, and that 'they have only a (sculling) oar and a rudder (at the stern)'. If the latter had not been something different from an oar, he would hardly have mentioned the two together in this way. The full meaning of the change described from one rudder to another according to the depth of water can be appreciated by a glance at the extremely deep rudders, acting also as centre-boards, which Korean ships have retained until the present day (Fig. 229).

A Japanese record, as graphic and personal as that of Hsü Ching [Xu Jing], permits us to take back his technical statements fully as far as the time of the book of Master Kuan [Guan] the geomancer. This is the *Nitto Guho Junrei Gyoki* (Record of a Pilgrimage to China in search of Buddhist Law) by the Japanese monk Ennin. What he says of the sea-going ships in which he travelled (and in which he was sometimes shipwrecked) shows that early in the ninth century axial rudders co-existed with steering-oars and stern-sweeps. Thus he describes a fierce storm, which drove the ship aground, 'where the corners of the rudder snapped in two places'. Soon afterwards the blade of the rudder stuck fast in the muddy sand. Here was no steering-oar, though there is probable mention of one when he was returning home a year later in another Japanese ship. He saw the sun setting 'in the middle of the great oar'. In the next breath, however he tells us that the moon set 'behind the stern rudder-

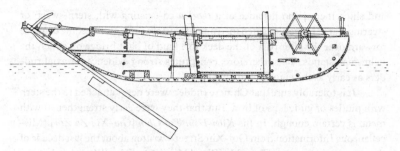

Fig. 229. Longitudinal section of a medium-sized (about 17 metre) fishing
vessel of traditional type as constructed on the small havens on Kanghwa
Island off the west coast of Korea. The extremely elongated rudder, which acts
as a centre-board when fully lowered, slides up and down in jaws, and its post
has a series of sockets into any one of which the tiller may be placed to suit the
helmsman's convenience. There are only two bulkheads but the transom bow
and stern are extremely blunt. The two masts carry very tall Chinese lug-sails
trimmed only slightly asymmetric. The size of the anchor winch is also
notable. (After drawings from H.H. Underwood).

house', so it would seem that an auxiliary steering-oar was carried as well as the
rudder with its tiller worked in the rudder-house (*tho-lou* [tou-lou]). This is
just what we have seen could happen with medieval ships, both European and
Chinese. Finally, a couple of months later, when Ennin was staying ashore
with Korean monks, a storm hit the same ship so that it was 'blown on to rough
rocks, and the rudder board broken off'. Thus all the eye-witness testimony of
Ennin vividly confirms the conclusions that have been drawn from the words
of Master Kuan and Than Chhiao [Tan Qiao].

   We turn now to how the steering device was fixed and of what material it
was made. The *Yü Phien* [*Yu Pian*] (Jade Page Dictionary) compiled about
543 says of the *tho* [*tuo*] that it is 'a piece of timber for regulating (the direction
of) a ship. It is set (*shê* [*she*]) (設) at the stern'. While many of the earlier
references refer to the stern, the verb used here implies something very
permanently established there, more so, perhaps, than would be warranted by
the lashings of a steering-oar. However, this is only a hint. The *Thang Yü Lin*
[*Tang Yu Lin*] (Miscellanea of the Thang [Tang] Dynasty) has to be taken
more seriously, for referring to about 780 it speaks of the ('rudder tower'), i.e.
the stern gallery or extension of the poop. This is the term always later used for
the projecting after-castle in or on which the helmsman stood (and still stands)
to work the tiller, and which also contains the winches or other arrangements
by which the rudder is raised and lowered (Fig. 227). So far as we know this
is the first appearance of the expression. Its significance cannot well be
overlooked, for with steering-oars there is no need of a *tho-lou*, indeed it would
be in the way, and among all the hundreds of existing types of Chinese boats

and ships there is no instance of a *tho-lou* co-existing with stern-sweep or steering-oars. These elongated instruments, where present, run some distance forward over the after part of the deck to a kind of light bridge on which the helmsman stands. This, therefore, constitutes strong evidence for axial rudders as early as the eighth century AD.

It is tolerably sure that Chinese rudders were never attached to the stern with pintles or gudgeons of iron. But that they were early strengthened with metal is certain enough. In his *Kuei-Hsin Tsa Chih* [*Gui-Xin Za Zhi*] (Miscellaneous Information from Gui-Xin Street), written about the last decade of the thirteenth century, Chou Mi [Zhou Mi] wrote 'The (best) wood at the bottom of the rudder is called the "iron corner". Sometimes the wood called *wu-lan-mu* from Chhinchow [Qinzhou] is used.' In another place when describing a storm at sea he said '. . . The rudder-post and the 'iron corner' made a terrible noise, "ya-ya, ya-ya", and seemed to be on the point of breaking. . .'. The ironmasters of this time would have been perfectly capable of arranging pintles and gudgeons if they had been required, but Chinese seamen always preferred to have the rudder movable up and down in guides; no doubt at the lowest position this greatly improved the windward sailing quality of their ships. Moreover, when sailing in heavy monsoon weather it was good to be able to keep the rudder down so as to protect it from breaking seas. And of course it was essential to raise it when beaching or sailing in the shallow waters of estuaries and the Yellow Sea.

There remains the question of the material of which the *tho* [*tou*] was made. The use of special wood seems to go far back in history. In the *San Fu Huang Thu* [*San Fu Huang Tu*] (Description of the Three Cities of the Metropolitan Area), which is probably late third century AD, there is a story which describes how the emperor ordered magnolia wood for the rudder. At the other end of history, we may remember the details given by Sung Ying-Hsing [Song Ying-Xing] fourteen centuries later about iron-wood for rudders (p. 83 above). Intermediate in date (1187) is the *Ling Wai Tai Ta* [*Ling Wai Dai Da*] (Information of What is Beyond the Passes) of Chou Chhü-Fei [Zhou Qu-Fei]. He gives an interesting description of special South Chinese woods much sought after for rudders, and says

> The Chhinchow [Qinzhuo] coastal mountains have strange woods, of which there are two remarkable kinds. One is the *tzu-ching-mu* [*zi-jing-mu*] (purple thorn tree) as hard as iron and stone, in colour red as cosmetic paint, and straight-grained; as large in girth as two men's reach, and when used for roof beams will last for centuries. The other kind is the *wu-lan-mu*; it is used for the rudders (*tho* [*tuo*]) of large ships, for which it is the finest thing in the world. Foreign ships are as big as a large house. They sail the Southern Seas for several tens of

thousands of *li*, and thousands or hundreds of lives depend upon one rudder. Other varieties of (timber for) rudders are not more than 9 metres in length, and are good enough for junks with a capacity of 10 000 bushels, but these foreign ships carry several times this amount, and might break in two if they encountered storms on the deep sea. But this Chinnchow timber is dense and tough, comes in 15-metre lengths and is not affected by the anger of the winds or waves....

This '*wu-lan-mu*' wood was probably the same as the iron-wood of later times, but the passage is rather ambiguous about the kind of rudder for which it was destined. Chou Chhü-Fei has just told us of wood for the *tho* [*tou*] of the 'foreign ships' of the Southern Seas many metres long. It was unfortunate that he expressed himself just in this way, for if he meant only 6 or 9 metres the length would not be at all excessive for a true axial rudder of substantial size with rudder-post included, but if he meant 20 or 24 metres, longer than even the lengths of wood from Chhinchow and elsewhere, he could only have been referring to a great stern-sweep.

(ii) Pictorial and archaeological evidence

The foregoing discussion, based solely on the words of ancient and medieval writers, was first sketched out in 1948. It has to be completed by the evidence from pictorial representations which have survived, and finally by archaeological evidence. The former, though of great interest, proved incapable of taking us much further than the point to which the texts had already led us, but the latter, just a decade later, turned out to be quite decisive, settling the matter in a way more radical than anyone had anticipated. It has shown that the other lines of argument were fully justified, and has proved what they could only surmise.

Let us proceed as before, starting from the earliest times and moving forwards. We may then also work backwards from the most reliable Chinese pictures of ships which date from times later than the European appearance of the stern-post rudder. In this way we shall be focusing, as it were in a microscope, both from below and from above. The pictorial counterparts of the Han and San Kuo [San Guo] passages we have referred to are of course the familar relief carvings of the tomb-shrines, in which small boats with steering-paddles are frequently seen (Fig. 230). There are also the boats, large and small, depicted on Indo-Chinese bronze drums; these show steering-oars being used. This takes us down to the third century AD. From then until the end of the Thang [Tang] (906), we are mainly dependant upon the paintings and carvings of the Chinese Buddhists, such as the frescoes of the Tunhuang [Dunhuang] cave-temples (Fig. 231) and the steles of the Liu Chhao [Liu Que]

224

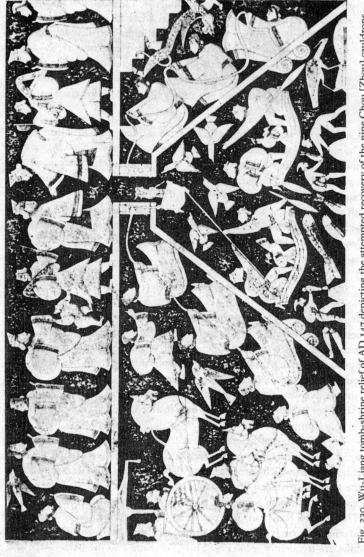

Fig. 230. Wu-Liang tomb-shrine relief of AD 147 depicting the attempted recovery of the nine Chou [Zhou] cauldrons from the river. These cauldrons were supposed to have come from the legendary Hsia [Xia] kingdom, (and to have been cast with pictures or maps illustrating various places or regions and the strange things found there. (See volume 2 of this abridgement, p. 241). Our interest here in the relief is that it shows small boats with steering-paddles.

Fig. 231. The largest of the ships in the Tunhuang [Dunhuang] cave-temple frescoes. The Buddhist Ship of Faith sails from the shores of illusion in the foreground (the upright oblongs are inscription-bearing cartouches) to the Paradise of Amida. The square-ended bow and stern of this early Thang [Tang] (seventh-century AD) ship are indeed very Chinese, with characteristic aft prolongation of the uppermost length-wise timbers. However the bellying square-sail is exceedingly un-Chinese, though suitable for a boat of the Ganges, and the latter would also have arrangements for working the quarter steering-paddles closely similar to those delineated here. Photograph Pelliot Collection, Museé Guimet, Paris.

period (Fig. 232). In these the steering-oar uniformly persists, even when the craft are seemingly quite large; there are no rudders.

What of the other end? Here we are troubled by the question of authenticity, for Chinese and Japanese artists did not always faithfully reproduce the technical detail in the paintings which they copied, and the number of authentic examples from the Sung [Song] is now relatively small. Nevertheless Yuan (Mongol) paintings of the thirteenth and fourteenth centuries always show rudders below the high curving sterns of the ships, and the Suzuki collection in

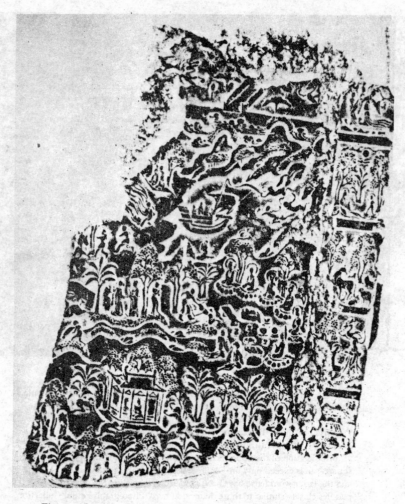

Fig. 232. Carving of a ship on a Buddhist stone stele of the Liu Sung [Liu
Song] or Liang dynasties (fifth or sixth centuries AD) from the Wan Fu Ssu
[Wan Fu Si] temple at Chhêngtu [Chengdu]. Photograph, Historical Museum,
Szechuan [Sichuan] University.

Japan has a Sung painting dating from before 1180 of two junks with finely
balanced rudders. There is also in Chinese possession a painting of 'Sailing
Ships making for Market by a Mountain Shore' which is placed at or before
1200..

With the Chinese ship carved so grandly by Cambodian artists on the

Fig. 233. Details of a slung and balanced rudder system of one of the cargo-boats in the twelfth-century painting 'Going up the River to the Capital after the Spring Festival'. (See also Fig. 200 for another section of this picture.)

walls of the royal city of Angkor Thom (Fig. 199) we are in a good position, for its date of 1185 is not in doubt. That it shows a Chinese merchant-ship is certain from many features (mat-and-batten sails with multiple sheets, grapnel anchor, etc.), but at the after part of the hull there is seen what at first sight appears to be an axial 'stern-post' rudder turned round so that it is facing forward along the quarter. But as it projects below the hull, acting as a centreboard in the usual way, and since the stern-gallery (from which it would be hoisted and lowered) juts out above it – all features highly characteristic of the Chinese slung rudder – there seems little doubt that it must be a true median rudder. Indeed, one can even see the head of the helmsman inside a *tho-lou* [*tou-lou*] at deck level.

Then as our next document we have the celebrated scroll-painting by Chang Tsê-Tuan [Zhang Ze-Duan] finished just before 1125 and entitled 'Going up the River to the capital, Khaifêng [Kaifeng], after the Spring Festival', with all its wonderful wealth of detail on the daily life of the people and their techniques. Here many slung and balanced rudders are depicted with the utmost clarity (Fig. 233). Since the date corresponds almost exactly with the literary evidence of the embassy to Korea we have once again the mutual confirmation of painting and text.

Lastly, if the existing Sung [Song] copies are to be trusted, we can find balanced rudders in the paintings of the famous artist Kuo Chung-Shu [Guo Zhong-Shu], who, in his 'Sailing on the River while the Sky is clearing after Snow' executed in 951, depicted two large junks with well-drawn balanced rudders. Between 950 and 1000 there is only one piece of pictorial evidence, a painting of a ship by Ku Khai-Chih [Gu Kai-Zhi] the famous artist of the second half of the fourth century AD. While it seems to embody certain archaic ideas of perspective characteristic of draughtsmanship at the time, the earliest extant version of it is probably a Sung copy of the eleventh or twelfth century. It is therefore impossible to be sure that the rudder-like object shown was really original.

Since 1958 all such doubts could be set at rest. For in the previous years excavations undertaken by the Kuangtung [Guangdong] Provincial Museum and Academia Sinica in the city of Canton, in connection with rebuilding operations, brought to light from a tomb of Hou Han date (first and second centuries AD) a magnificent pottery ship model which demonstrated the existence of the axial rudder already at that time. Before these discoveries the tomb ship models recovered in modern archaeological research had all been of the Warring States or Early Han period (fourth to first centuries BC), and all had evidence of steering-oars. Now however the pottery model, nearly 60 centimetres long, was found to be equipped in very modern style. We need not recapitulate the description already given (p. 106), suffice it to say that, as Fig. 234 shows, deck-houses cover most of the beam, which is surrounded on both sides by a poling gallery. The stern extends aft a considerable distance beyond the last transom-bulkhead in the form of an after-gallery (in fact a *tho-lou* [*tuo-lou*]), the floor of which is formed by a criss-cross of timbers through which the rudder-post descends into the water. This is seen particularly well in Fig. 235, a photograph taken from astern. The true rudder is indeed present, shaped like a trapezium as would be expected, and having no resemblance to a steering-oar, but most clearly exemplifying the remark about '2.5 metres of timber' which Than Chhiao [Tan Qiao] was to make nearly a thousand years later. Most gratifyingly, its shoulder is pierced by a hole exactly where the suspending tackle should be. Possibly the original state of the model, made no doubt for some wealthy merchant-venturer and ship-owner of Han Canton, incorporated all the tackle by which the rudder was secured, but the little cables long ago rotted away and we can only guess how it was done. It is very noteworthy also that the rudder is distinctly balanced, about one third of the breadth of the blade being in front of the axis of the post.

Thus on the main issue guesswork is ended, and we have positive demonstation that by the first century AD the true median rudder had come into being. How strange it is that this was just the time, too, to which one may

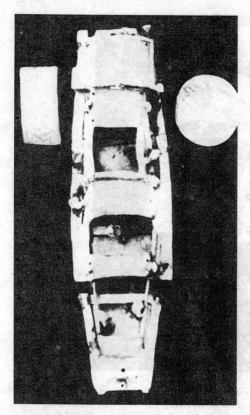

Fig. 234. The grey pottery tomb-model ship of the Later Han (see also Fig. 197) viewed from above. The bow is at the bottom of the picture. The removable roofing shows the cabins or holds, but no bulkheads are visible. The wide poling galleries break off only in one place, the probable position of the mast, but no tabernacling is provided. Photograph Canton Museum.

trace back the first beginnings of the magnetic compass. How strange also that though the latter was much slower in its development than the steering mechanism, it appeared in Europe at the same time as the rudder, just a millennium later. The only difference between them is that the compass in the West is recorded first from the Mediterranean, the axial rudder first from Northen European waters.

(iii) Transmissions and origins

About the transmission of the technique (and surely such there was) very little can be said. It seems overwhelmingly likely that an invention of this kind would have come round by way of mariners' contacts in the South Asian

Fig. 235. Stern view of the Later Han tomb-model ship of Figs. 197 and 234. The attachment of the axial rudder, with its eye for the slinging tackle, between the timbers of the floor of the after-gallery, is well seen.

seas, though it is not impossible that a Chinese artisan who had built ships for the Liao dynasty handed on certain ideas to Russian merchant-shipwrights trading to Sinkiang [Xinkiang] between 1120 and 1160. This might explain the region in which the rudder first manifests itself in European culture, but support from Russian sources is so far lacking.

At the same time the Islamic world offers more light (though not very much) for the travels of the rudder than it does for the mariner's compass. A famous illustration in a Baghdad manuscript of 1237 shows an axial rudder on a sewn ship. The rudder seems to be provided with some form of lateral control, but no illustrative material of the required time, a century earlier, has so far come to light. However, a description of a related device was given by Lecomte, and for the past century and a half European observers have described elaborate tackle-controlled rudders on many types of Arabic sailing-

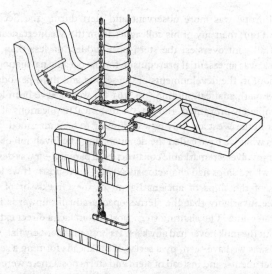

Fig. 236. Drawing of the rudder of a Fuchow [Fuzhou] timber freighter (after G.R.G. Worcester). The rudder, weighing some 4–8 tonnes, measures 9.9 metres in height and 3.5 metres in width; it is hoisted and lowered by a bight of chain passing through a sheave in the blade, both parts being wound round the barrel of the windlass above. The rudder post, 4.6 metres long, is iron-bound at intervals of 30 centimetres throughout.

ship. Hence considerable interest attaches to a description by al-Muqqaddasi in 935. Describing a passage in the Red Sea he says:

> The shipmaster takes his stand ... and steadily looks into the sea. Two boys are likewise posted on his right hand and on his left. On espying a rock he at once calls to either of the boys to give notice of it to the helmsman by a loud cry. The latter, on hearing the call, pulls one or other of two ropes which he holds in his hand to the right or the left, according to the directions. If these precautions are not taken, the ships stands in danger of being wrecked against the rocks.

It seems almost impossible that this description could refer to lanyards attached to steering-oars, but on the contrary it would closely agree with the tackle-controlled axial rudders which have lasted in use in Arab waters to the present day; in this case we have to conclude that the Chinese invention had already been introduced in the Arab culture-area before the end of the tenth century. From all we know of Arab trade in the Eastern Seas, this would not be at all extraordinary. But the transition from the Muslims to Northern Europe remains at first sight more difficult to understand. Perhaps some sea-captain from Northern Europe was more observant and alert during the Second Crusade (1145 to 1149) than any of his colleagues from the Mediterranean.

In spite of all controversies, the stern-post rudder, no less than the mariner's compass, was an essential prerequisite for the oceanic navigation of large ships. Without it, the developments of the second and third periods of measurement and mathematical piloting (pp. 161 ff.) would have been long delayed if not completely prevented. Indeed, the historical implications of the stern-post rudder in the West are only now beginning to be understood.

Now comes what we have called the dénouement. The invention of the stern-post rudder involves a remarkable constructional paradox – it was developed by a people whose ships had characteristically no stern-posts. If we look again at pictures of the ships of ancient Egypt, of the Greeks, or of the Norsemen, we see invariably that the stern sloped gradually upwards in a curve from the water-line. The slanting stern-post was in fact a direct extension of the keel. But the junk never had any keel. Its bottom, relatively flat, was joined to the sides, as we have seen, by a series of bulkheads forming a set of water-tight compartments, and instead of stem and stern-posts there were flat transom ends. Now the bulkhead build provided the Chinese shipwrights with the essential vertical members to which the post of the axial rudder could conveniently be attached, not necessarily the aftermost transom but perhaps one or two bulkheads forward of it. This principle held good from the smallest to the largest sailing-ships. It might be called that of the 'invisible stern-post'. Of course, in later times, rudders were fashioned in curving shape so as to fit various kinds of curving stern-posts, but our argument suggests that the difficulty of doing this was one of the chief factors which prevented any earlier development of the invention in the West.

The bulkhead-attached rudder-posts can be seen clearly in many Sung [Song] pictures (Figs. 225 & 233), as well as in drawings of contemporary Chinese craft. The Cantonese ship model of the Han does not show the vertical nature of the attachment so well, but this may be partly because we do not know exactly how it slung its rudder – in any case the latter speaks for itself. Alternatively we may be seeing the axial rudder in a formative state, after it had acquired its very particular shape and just before it had found its home on the vertical bulkhead timbers. For it is noteworthy that although the Canton ship model has bulkheads its lines are rather like those of a punt, sloping gradually to the water-line at bow and stern; and only when more upright and blunt-ended sea-going forms developed from this would the invitation to verticality really have asserted itself – with the additional advantage of the centre-board function of the rudder in its lowered position. To sum up the matter, we can not only feel sure that the stern-post rudder originated in the Chinese culture-area at the beginning of our era, but we can form a pretty good idea of just why it did so.

Fig. 237. The rudder of a Ma-yang-tzu [Ma-Yang-zi] river junk of the Yangtze [Yangzi] seen in a shipyard on the Upper Yangtze. This beautiful shape, fitted on the right to the curve of the hull, is the Chinese balanced rudder *par excellence*.

BALANCED AND FENESTRATED RUDDERS

The civilisation (so often miscalled 'static') which initiated axial rudders also gave them a far-reaching development. From time to time we have had occasion to mention the 'balancing' of rudders. People generally think of the rudder as an object in which the whole of the blade or flat part lies behind the post. But many large modern ships, on the contrary, have rudders in which there is a flat portion in front of the post as well, and this construction is termed 'balanced'. Such an arrangement not only balances the weight on the bearings but also eases the work of the helmsman and the steering-power that may assist him, since the water exerts pressure in his favour on the forward portion. Balanced rudders are common on many types of Chinese river-junk (Fig. 237), though in their simple forms they are unsuitable for sea-going vessels. Although we have not been able to find any specific literary references to them there is now no room for doubt that they go back to the earliest stages of the invention in China. Indeed it seems quite likely that the balanced axial rudder was the first to evolve, for placing a steering-paddle in an upright mid-position against or near the aftermost bulkhead would lead directly to it.

Europeans were very slow, generally speaking, to adopt the principle, perhaps because they were mainly interested in sea-going ships, and until iron construction afforded ways of securing balanced rudders thoroughly (e.g. by pivoting the base of the post) they were not very feasible. An 'Equipollent Rudder' was among the inventions of Lord Stanhope about 1790, and the

Fig. 238. The stern of a Hongkong fishing vessel in dry dock, showing the fenestrated rudder characteristic of many types of Chinese ship. The figures of the workmen give the scale, and another fenestrated rudder being re-bladed is seen on the right. From the Waters Collection, National Maritime Museum, Greenwich.

matter was pressed forward in 1819, whilst one of the earliest ships with a modern balanced rudder was the *Great Eastern* of 1843. The strangest aspect of the situation was that one of the two oldest representations of rudders of any kind in Europe, that on the font at Winchester, shows what looks like a balanced rudder. Could this conceivably have any significance with regard to the transmission?

The balanced rudder is also traditionally at home in India, especially on the Ganges, where certain classes of boats, the *ulak* and the *patela*, for instance, are equipped with striking triangular forms. These seem more primitive than the Chinese ones however, because entirely symmetrical fore and aft, and therefore less efficient. One hesitates whether to look to China or to ancient Egypt as the main source of influence on the rudders of these vessels. But perhaps they are best regarded as very exaggerated Egyptian steering-paddles, all the more so because, though vertical, they are generally fixed to the curved part of the vessel's stern as quarter-rudders and not to the stern-post at all.

Perhaps the most remarkable of these inventions was the fenestrated rudder. When European sailors first frequented Chinese waters, they were surprised to see some junk rudders riddled with holes. No doubt they found it difficult to believe that this had been designedly done. Such fenestrations, generally diamond-shaped and cut out at the edges of the planks, ease the steering by reducing the pressure against which the tiller has to act, and minimise the drag on the ship caused by turbulence to the water flow past the rudder. But as water is in fact viscous, the efficiency of the rudder is very little impaired. Fig. 238 shows the stern of a Hongkong fishing-junk in dry dock with a fenestrated rudder. The device was probably merely the result of observing knotty wood or damaged gear, though it is not at all too fanciful to suppose that some Taoist [Daoist] sailor, finding that his work was eased and that his ship sailed better, was fully content to follow the principle of *wu wei* (no action contrary to Nature), and letting well alone, recommended the arrangement to his friends. The fenestrated rudder has been widely adopted in modern iron ships during the present century, having been brought to the attention of European marine engineers in 1901. Indeed, it may even have helped to stimulate the important invention of anti-stalling slots in the wings of aircraft.

7

Techniques of peace and war afloat

ANCHORS, MOORINGS, DOCKS AND LIGHTHOUSES

Anchors and moorings

Much has been written on the history of the anchor, an essential device which goes back to prehistoric times. The ancient Egyptians used heavy stones combined with hooked branches to form grapnels, but metal hooks were coming into use already in the European Tène Age (sometime about 400 BC). There exists a bronze age anchor about contemporary with Homeric mentions, and from about 500 BC onwards the anchor of the Mediterranean peoples had attained approximately the familiar form, as is known from many coins on which it appears. The stock, however, was absent in the earlier periods. That parallel forms occurred in China can be deduced from the names

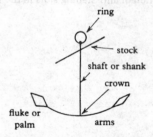

employed. The stone-weighted grapnel was called *ting* [*ding*] (矴), sometimes written (碇). It was simply made by binding from one to four forked branches together with a piece of stone, and this lasted long in use; we saw it on the Bayon junk of 1185 (Fig. 199). When the use of metal hooks was introduced, these words were replaced by *mao* (or *miao*) (錨), combining the idea of shoots of plants or the claw of a cat with the metal radical. The *Yü Phien* [*Yu Pian*] (Jade Page Dictionary) of AD 543 is the first in which the word *mao* appears, and if this may be taken as evidence that metal anchors were not used much earlier, their introduction in China would have occurred later than in the West.

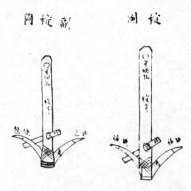

Fig. 239. Two adze anchors from the Fukien [Fujian] Shipbuilding manuscript, a larger (*ting* [*ding*]) and a smaller (*fu ting* [*fu ding*]). The tips of the arms are shod with iron.

Nevertheless the Chinese made a contribution of some importance to the development of the anchor. The most characteristic of their forms is the adze anchor, i.e. an anchor made in such a way that the arms form an acute angle with the shaft (about 40° or less) at the crown instead of diverging from it at a right angle or in an arc. Such forms were not unknown in Europe (the Roman temple-ships had them), but the Chinese passed the stock across the shaft not at the ring end but near the crown end (also of course at right-angles to the plane of the arms). This serves the purpose of canting the anchor and ensuring that the arms bite, but it has also the great advantage of being almost non-fouling. The efficiency of this device was often praised by European nautical writers, and during the nineteenth century it was several times 're-invented', with the additon of hinged stocks, so that some of the modern 'stockless' types derive originally from the Chinese rather than the Graeco-Roman form. We can trace the adze anchor back through the Fukienese [Fujianese] Shipbuilding manuscript (Fig. 239) not only to the *Wu Pei Chih* [*Wu Bei Zhi*] (Treatise on Armament Technology) of 1628 (Fig. 240) but to the first century AD in the Cantonese tomb-model ship (Fig. 241).

Anchor windlasses (i.e. vertically mounted drums) are mentioned several times in the *Kao-Li Thu Ching* [*Gao-Li Tu Jing*] (Illustrated Record of an Embassy to Korea) of 1124, and were carried on both the Korean and Chinese ships. Those on the larger Chinese 'retainer ships' which carried the personnel of the embassy, Hsü Ching [Xu Jing] tells us, had a rope which was as thick as the rafters of a house, and was made of twisted *thêng* [*teng*] (some kind of vine or liana) 150 metres long. A slightly later depiction of an anchor windlass can be seen on the Bayon junk (Fig. 199). Sea-anchors are also referred to by Hsü Ching under the name of *yu ting* [*you ding*], but he erred in saying they were just like ordinary anchors, though he knew they were used in

Fig. 240. An adze anchor from the *Treatise on Armament Technology*, printed
in 1628. The wooden label says: 'Anchor belonging to Official Ship No.
so-and-so.'

stormy weather at sea. In fact, they must then have been what they are now,
large bamboo baskets, in the use of which no sailors are more skilled than the
Chinese.

As for the seventeenth century, Sung Ying-Hsing [Song Ying-Xing]
has this to say about the anchors of the inland grain-transport freighters:

> Anchors of iron are dropped into the water to moor the ships; a
> grain-boat usually has five or six. The heaviest weighs about 500
> catties (about 300 kg) and is called a 'watchdog'. In addition, two
> small ones are slung both at bow and stern. When a ship in midstream
> meets with too strong an adverse breeze, so that she cannot go
> forward, and also cannot tie up anywhere, [or where the river-bed
> near shore is rocky instead of sandy, and one cannot approach the
> shore, then one must anchor in deep water] letting go the hook so that
> it sinks (quickly) to the bottom. The hawser of the anchor is wound
> round the bollards on the deck (and made fast to them). When the
> flukes of the anchor touch the mud and sand of the bottom they dig in
> and hold securely. The 'watchdog' is resorted to only when danger is
> imminent; its hawser is called 'ship's-self' (*pên shen* [*ben shen*]) in
> order to indicate its importance. Or again, when the boat is under way
> in company and seems likely to collide with another vessel ahead
> which has had to slow down, the stern anchors are smartly lowered
> into the water to check the speed. As soon as the wind abates, the
> anchors are hoisted by means of a winch.

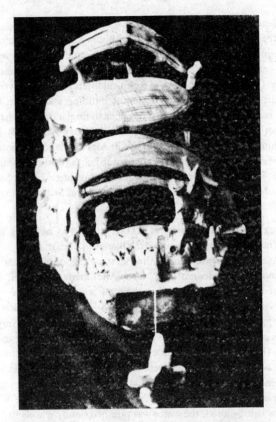

Fig. 241. Bow view of the Later Han tomb-model boat of grey pottery. (See also Figs. 197, 234 and 235). The anchor is seen attached to its bollard, and just behind this is an ornamental screen which recalls the 'prow-yokes' of Indo-Chinese craft. Lateral projections support narrow outboard decking and bits or bumkins (projecting booms). The transom stem is well seen. Photograph Canton Museum.

A very different form of mooring is commonly found in use by the smaller river-junks and sampans; it consists of a trunk or tube built into one or more of the compartments of the vessel. Through this a weighted pole is driven down into the mud of the lake or river bottom. Known as the 'water-eye' (*shui yen* [*shui yan*]) this device has the advantage of continuous adaption to fluctuating water-levels. In China this pole-setting or 'stick-in the-mud' anchor is at least as old as the Sung [Song], for paintings of that period show it, but it is ancient in many parts of the world from New Guinea to sixth-century Holland. The principle is still used in modern dredgers.

Harbours and docks

To study the construction and layout of harbours and docks in Chinese history would require a whole chapter to itself, all the more difficult to write in that we have come across no studies of the subject either by Chinese or Western scholars. One question of marked technical interest may however be raised, namely the development of dry docks for building and repair. The European aspect is a little obscure. The first dry dock in England, and possibly in the Western world, was that made at Portsmouth for Henry VII in 1495, but some scholars claim that such docks were in use in Alexandria (third century BC), though more recent research has brought forward nothing to substantiate this. In any case we have excellent evidence for the invention in the Sung [Song] period, with a circumstantial account from the pen of Shen Kua [Shen Gua]:

> At the beginning of the dynasty (*c.* AD 965) the two Chê [Zhe] provinces (now Chekiang and southern Chiangsu [Zhejian and southern Jiangsu] presented (to the throne) two dragon ships, each more than 60 metres in length. ... ... After many years, their hulls decayed and needed repairs, but the work was impossible as long as they were afloat. So in the Hsi-Ning [Xi-Ning] reign period (1068 to 1077) a palace official Huang Huai-Hsin [Huang Huai-Xin] suggested a plan. A large basin was excavated at the north end of the Chin-ming [Jin-ming] Lake capable of containing the dragon ships, and in it heavy cross-wise beams were laid down upon a foundation of pillars. Then (a breach was made) so that the basin quickly filled with water, after which the ships were towed in above the beams. The (breach now being closed) the water was pumped out by wheels so that the ships rested quite in the air. When the repairs were complete, the water was let in again, so that the ships were afloat once more (and could leave the dock). Finally the beams and pillars were taken away, and the whole basin covered with a great roof so as to form a hangar in which the ships could be protected from the elements and avoid the damage caused by undue exposure.

Apparently the Chinese emperor's dock had no swinging gates any more than that of Henry VII, but he did have four centuries priority.

Lighthouses

While speaking of harbours, havens and anchorages, a word may be said about the lighthouses which help one to get there. They are perhaps less prominent in Chinese than in ancient European literature, as might be expected from the relatively greater importance of maritime navigation in the West. References to beacons in Chinese encyclopaedias almost always refer to lights on hills or forts for military or other governmental signalling. There was nothing similar to the Pharos of Alexandria, built in 270 BC by Sostratus of Cnidus for Ptolemy Philadelphus, which probably reached a height of 45 metres and much of which was still standing in the thirteenth century. Though coastwise and lakeside lights must have been used on a small scale in China, literary references generally speak of them in connection with foreign parts. Thus Chia Tan [Jia Dan] the geographer, writing between 785 and 805, says in his description of the sea route between Canton and the Persian Gulf, referring to some place near the latter's mouth, 'the people of the Lo-Ho-I [Lo-He-Yi] country have set up ornamental pillars in the sea, on which at night they place torches so that people travelling on board ships shall not go astray'. Independent confirmation of lighthouses in the Persian Gulf for a century later is available in Arabic authors. It is perhaps of interest to read what a Chinese writer said of the Alexandrian Pharos in 1225:

> The country of O-Kên-Tho [O-Gen-Tuo] (Alexandria) belongs to Egypt. According to tradition, in olden times a stranger, Chhu-Ko-Ni [Chu-Ge-Ni] by name, built on the shore of the sea a great pagoda, underneath which the earth was excavated to make two rooms, well connected and thoroughly hidden. In one vault was stored grain, and in the other arms. The tower was 60 metres high. Four horses abreast could ascend (by a winding ramp) to two-thirds of its height. Below the tower, in the middle, there was a well of great size connected by a tunnel with the great river. To protect this pagoda from foreign soldiers, the whole country guarded it against all enemies. In the upper and lower parts of it twenty thousand men could readily be stationed as a guard or to make sorties. At the summit there was an immense mirror, and if warships of other countries tried to make an attack, the mirror detected them beforehand, and the troops were ready to repel it. . . .

With this background, it is rather interesting that one of the most famous Chinese lighthouses was the minaret of a mosque in Canton. A description of the building was written about 1200. Rising to a height of 50 metres, it was called the 'Light Tower' apparently because a light was kept burning at the top to guide shipping. It was first built in the Thang [Tang] by foreigners, we are told, and had a spiral staircase inside it. Every year in the fifth and sixth months the Arab foreigners used to assemble to scan the estuary for their sea-going barques, and then at the fifth drum ascended the tower to shout prayers for favourable winds. In 1468 the Imperial Censor Hang Yung [Han Yong] caused the minaret to be repaired, and arranged it for sending official messages, presumably by lights. However, Buddhist pagodas also served occasionally as lighthouses. The *Hang-chou Fu Chih [Hang-zhou Fu Zhi]*

Fig. 242. Salt boats being tracked upstream, an illustration from the *Szechuan Yen Fa Chih* [*Sichuan Yan Fa Zhi*] (Memorials of the Salt Industry of Szechuan and its control) (1882). The whirlpool adds to the drama of the drawing, but the rivers of Szechuan have them, as Joseph Needham experienced in 1943.

(Gazeteer [Historical Topography] of Hangchow [Hangzhou]) of 1686 says that the Liu-Ho Tha [Liu-He Ta] on the Chhien-thang [Quiantang] River was equipped with a permanent light from the early Sung [Song] onwards to guide ships seeking their anchorages at night. Thus two at least of China's religions contributed something as the equivalent of the Brethren of Trinity House.

TOWING AND TRACKING

Mention has often been made above of the tracking of junks up the rivers in China. For anyone who has lived near one of the great Szechuanese [Sichuanese] rivers, the Chialing [Jialing] at Chungking [Chongqing] for example, the cries of the trackers, and the sound of the drums which give them the time, remain unforgettable memories. No natural difficulties defeated the Chinese boatmen, and towing galleries were cut even into the precipitous rock faces along the gorges of the Yangtze [Yangzi]. Teams of as many as a hundred men may be employed in the most difficult places. Figure 242 shows salt-boats being tracked upstream in Szechuan [Sichuan]. The paintings of Hsia Kuei [Xia Gui] (1180 to 1230) are often cited as evidence of tracking before Marco

Polo's description, but we have much earlier representations, such as those among the Tunhuang [Dunhunag] frescoes.

In tracking, the cable is secured to a crossbeam aft, whence it passes to the mast and runs through a cast-iron snatch-block (a block with a hinged plate on one side to take a rope without threading). This can be raised or lowered by ropes and tackles; normally it hangs at about one third of the height of the mast, but when overtaking another boat it is hoisted to the masthead. A bamboo cable is used for tracking and one of its remarkable qualities is that while hempen ropes lose some 25 % of their strength when wet, this plaited bamboo undergoes on saturation with water an increase of 20 % in tensile strength. Test have shown that a cable of 3.75 centimetres diameter will take a load of nearly 5 tonnes when dry and about 6 tonnes when wet. The cloth-band harness worn by the trackers may well be related to the development of an efficient harness for draught animals, which will be discussed in a later volume of this abridgement.

CAULKING, HULL-SHEATHING AND PUMPS

The means used by traditional Chinese shipwrights and sailors to render hulls watertight have been referred to from time to time in what we have discussed previously. Generally speaking the classical mixture for caulking was tow mixed with tung oil and lime, while additional security was obtained in old ships by nailing on season after season fresh layers of strakes to increase the hull's thickness. The best varieties of marine drying putty were, however, of more complex composition, especially notable being the addition of a proportion of soya-bean oil.

Protection against enemies other than water was a rather different matter. Sheathing hulls against invasion by the shipworm (*Teredo* – a bivalve mollusc) and other pests, or the growth of sessile marine organisms like barnacles, and the protection of upper works against the attacks of the enemy in battle, go naturally together. But metal plates were used for the former purpose long before their employment for the latter. The Romans tried sheathing ship bottoms with lead, as in a galley of Trajan on Lake Ricio, and in the temple-ships of Lake Nemi. This, however, was exceptional and no examples of sheathing are known from the Middle Ages. Lead was again tried in Europe about 1525, but soon abandoned in favour of sheathing with a layer of boards often with horse-hair packing; then from 1758 copper plates began to be employed. Yet Chinese writings of the early fourth century AD refer to the covering of junk bottoms with copper. Thus the *Shih I Chi* [*Shi Yi Ji*] (Memoirs on Neglected Matters) of about 370, referring to an embassy from the Jan-Chhiu [Ran-Qiu] kingdom during the legendary reign of Chhêng Wang [Cheng Wang], says: 'Floating on the seething seas, the ambassadors came on a boat which had copper (or bronze) (plates) attached to its bottom, so that the crocodiles and dragons could not come near it.' Here the defence against organic life is clearly mentioned, and the passage would seem to prove that the idea, at least, existed in Wang Chia's [Wang Jia's] time. Another Chin [Jin] book, the *Chiao-Chou Chi* [*Zhao-zhou Ji*] (Record of Chiaochow [Zhaozhou] (District)) says that at Anting a copper or bronze boat which had been built for the King of Yüeh [Yue] lay for a long time buried in the sand, where it could be seen at low tide.

It has now been shown that stories of metal boats occur abundantly in the early Chinese literature of folklore and legend. They are particularly common in South China and Annam, where they often form part of the exploits of the Han general, Ma Yuan, who restored the far south to Chinese allegiance in the campaign of AD 42 to 44. The bronze or copper boats of which people see the vestiges are thus associated with the setting up of bronze columns to mark the southern limits of the empire, the casting of bronze oxen as landmarks, and the building of canals to shorten sea voyages or make them more safe. The evidence comes from texts dating from all periods between the third and ninth centuries AD, but the only one which specifically mentions the bottom of a ship is the early one just quoted. Although it is quite possible, as sinologists tend to think, that the idea of using metal in the construction of boats was purely magical and imaginary in origin, it is at any rate equally possible that some southern group of shipwrights in those ages had the services of smiths who beat metal into plates fit for nailing to the hulls of their craft to protect the timbers. If so, the copper-bottomed junks of the eighteenth century derived from an ingenious tradition, and not from the Lake of Nemi in the Far West. We even hear of iron boats. In or before the Sung [Song], a book of unknown authorship, the *Hua Shan Chi* [*Hua Shan Ji*] (Record of Mount Hua), speaks of a derelict iron boat beside a mountain lake. This was doubtless a further echo of the same legend, or the same technique. But iron armour for warships was no legend, as we shall soon see.

The machines used to keep the hulls of ships dry when afloat have been little studied. The large ship built about 225 BC by Hieron of Syracuse was said to have been fitted with an Archimedean screw, worked by one man, for pumping out the bilges, but the account by Athenaeus (second century BC) seems somewhat fabulous.

After the end of the sixteenth century the Chinese used piston pumps, as Europeans did earlier. But under early technological conditions such machines were probably much less effective than chain pumps, and indeed we find that Westerners who came in contact with Chinese shipping at this time greatly admired the methods employed. One of the earliest descriptions is by Gaspar da Cruz, a Portuguese Dominican who visited China for a few months in 1556, followed in 1585 by Gonzales de Mendoza who also appreciated the

Chinese technique and wrote:

> The pumpes which they have in their shippes are much different
> from ours, and are farre better; they make them of many peeces, with
> a wheele to draw water, which wheele is set along the ship's side
> within, wherewith they doe easily clense their shippes – for that one
> man alone going in the wheele, doth in a quarter of an hour cleanse a
> great shippe, although she leak very much....

A century later it was popularised by the Dutch scholar Isaac Vossius, and
still thought noteworthy at the beginning of the nineteenth century. As we lack
clear Chinese descriptions, it is a little difficult to be sure about the type of
chain pump used. The description by da Cruz points unmistakably to an
inclined square-pallet chain pump of true Chinese type: Before the middle of
the sixteenth century, then, in this connection at any rate, the Chinese had not
been much inhibited by the absence of the piston pump. The two machines, in
fact, travelled in opposite directions at this time, chain pumps being adopted
on Western ships, and piston pumps attracting attention in China.

There was another purpose for which pumps were valued at sea, namely
for extinguishing fires caused by incendiary enemy attack. This subject has
already been mentioned from time to time in passing, but it needs a word more
here. In the *Sung Hui Yao Kao* [*Song Hui Yao Gao*] (Drafts for the *History of
Administrative Statutes of the Sung* [*Song*] *Dynasty*) of 1129 there is a reference, among other things, to equipping warships with fire-protection devices, and elsewhere there has been talk of wet leather curtains for protection
against fire-arrows. Then about 1360, Su Thien-Chio [Su Tian-Jue] in his
Kuo Chhao Wên Lei [*Guo Chao Wen Lei*] (Classified Documents of the Present
Dynasty) speaks as follows concerning the Battle of Yai-shan between the
Sung [Song] and Yuan (Mongol) fleets which occurred early in 1279:

> Later (Chang) Hung-Fan [(Zhang) Hong-Fan], having captured a
> number of boats of the Tan-chia [Dan-jia] river-people, caused straw
> to be piled on them and soaked with oil; then, when the wind was
> favourable, he had them set adrift as fire-ships to burn the Sung
> [Song] fleet. But the Sung ships had previously been plastered all
> over with mud, and moreover countless numbers of 'water-tubes'
> (*shui thung* [*shui tong*]) were suspended over (their sides). So when the
> piles of blazing straw came near, they were pulled to pieces with
> (long) hooks, and extinguished with water. Thus none of the Sung
> ships was harmed.

Although, then, the exact nature of the system remains unclear, there must
have been tanks on the upper deck kept supplied by pumps and capable of
providing water for sprays and hoses.

设水采珠船

Fig. 243. Pearl divers at work, from the *The Exploitation of the Works of Nature* (1637). The caption says: 'The ship which carries the divers who go down into the sea to collect pearls.' Breathing-tubes and some kind of masks are in use, but in the corresponding picture of the Ming edition, only the latter are seen.

DIVING AND PEARLING

At one or two points already there has been mention of submersible craft, of course a late development in any civilisation. Here it may not be out of place to refer to efforts of earlier times to enable human divers to remain under water at considerable depths for as long as possible. China comes into the story in connection with pearling. In one of the later chapters of the *Thien Kung Khai Wu* [*Tian Gong Kai Wu*] (The Exploitation of the Works of Nature) of 1637, Sung Ying-Hsing [Song Ying-Xing] describes the pearl fisheries, which in his time (apart from foreign countries of the South Seas) were concentrated near Leichcw and Lienchow [Leizhou and Lianzhou] in southern Kuangtung [Guangdong], north and north-west of Hainan Island. He tells us that the divers, who belong to the Tan [Dan], and ancient southern people, worked over special pearl-oyster beds, using broad-beamed boats peculiar to themselves. They descended, he says, as much as 400 to 500 (Chinese) feet, though there seems some misunderstanding here. Such a depth would have been quite impossible without the use of a modern diving suit and helmet, which was not on the cards; it seems that 20 metres (70 feet) is the limit at which divers

底沉芭竹 珠採帆揚

Fig. 244. Li Chao-Thao's [Li Zhao-Tao's] dredge or drag-net for collecting
pearl oysters, from the *Exploitation of the Works of Nature* (1637). The caption
on the right says: 'Hoisting sail to collect pearls', and on the left: 'The bamboo
drag that goes down to the bottom.'

without respiratory aids can work, though they may go down 36 metres (120
feet) for very short periods. The Tan divers had a long rope from a winch
secured to their waists, and breathed through a curving pipe strengthened by
rings of tin, which was fastened over the face with a leather mask (Fig. 243). If
anything went wrong, they pulled on the rope as a signal and were quickly
hauled up, but many were unfortunate and 'found a tomb in the bellies of
fishes', while others died of cold after leaving the water. Apparently to relieve
these hardships, a Sung [Song] inventor, Li Chao-Thao [Li Zhao-Tao],
devised a kind of weighted dragnet with iron prongs like a plough, and a
hempen bag the mouth of which was held open to receive the oysters, so that
the pearl-fishermen could tow their dredge while under sail (Fig. 244). In
Sung Ying-Hsing's time, both methods were used. He adds that sometimes,
for tens of years at a time, there was a sort of close season, to allow the pearls to
grow.

The centre of the industry lay in the Lienchow [Lianzhou] region, the
old name of which was Ho-phu [He-pu] commandery, and here on the coast
among islands there were 'pearl lagoons' or indentations of the sea so famous
that at one time the whole district was known by that name. This wealth had

been exploited at least as far back as 111 BC, when the armies of Han Wu Ti [Han Wu Di] annexed the old kingdom of Yüeh [Yue], and the abundant production of pearls is recorded already in the *Chhieh Han Shu* [*Qian Han Shu*] (History of the Former Han Dynasty) (*c.* AD 100). Towards the end of the first century BC people from other parts of China made fortunes by organising the work of the pearl-divers. So also did successive governors, with the result that over-fishing produced scarcity, and it took the action of a wise and good man about AD 150 to retrieve the situation. Thus Mêng Chhang [Meng Chang] decreed a temporary cessation of pearling, and stands out as a successful exponent of nature-protection and fisheries conservancy. In char-acteristic Chinese fashion he later became the tutelary deity of the industry, and long afterwards Thao Pi [Tao Bi] (1017 to 1080) wrote an inscription for his temple:

> In bygone times good Governor Mêng,
> Loyal and honest, walked by this distant shore.
> He did not rob the wombs of the oysters,
> And the waters' depths abounded with returning pearls.

In the third century AD, after the end of the Han, the pearling districts became part of Wu State, and from this time dates what may be the earliest account of the divers' work. In his *Nan Chou I Wu Chih* [*Nan Zhou Yi Wu Zhi*] (Strange Things of the South), Wan Chen [Wan Zhen] wrote:

> There are people in Ho-phu [He-pu] who excel in swimming to search for pearls. When a boy is ten or more years of age he is instructed in pearl-diving. The officials forbid the folk to gather pearls (except for the government). But certain skilful robbers, crouching on the sea bottom, split open the oysters and get fine pearls, whereupon they swallow them and so come forth.

Smuggling was thus keeping pace with government control. And indeed from 228 the whole district had been renamed for a while Chu-kuan [Zhu-guan], i.e. (the domain of the) Director of Pearling.

Over the ages the industry was troubled from time to time by waves of Confucian austerity at court which injured all such luxury trades, and during the Thang [Tang] pearling was stopped several times. No such inhibitions weighed on Liu Chhang [Liu Chang], however, the last of the emperors of the Nan Han dynasty in the Wu Tai [Wu Dai] period, who stationed a whole division of soldiers near Lienchow [Lianzhou] and had them instructed in pearl-diving. But as soon as the Sung [Song] armies took Canton in 971, this use of troops was abolished.

One of the earliest texts which attributes to the Tan [Dan] people the greatest role in the pearling industry is the *Thieh Wei Shan Tshung Than* [*Tie*

Wei Shan Cong Tan] (Collected Conversations at Iron-Fence Mountain) written about 1115 by Tshai Thao [Cai Tao]. In a long and interesting passage he tells us that the fishermen arrange ten or more of their boats over the pearl beds in a ring, and let down on both sides mooring-cables attached to rocks which lie as anchors on the bottom. Then the Tan diver, having attached a small rope to his waist,

> ... takes a deep breath and plunges straight down from 3 to 30 metres, after which he leaves the mooring-cable and feels his way to collect the pearl-oysters (lit. pearl-mothers). After what seems only a few moments he urgently needs air, so he gives a big jerk to the waist-rope, and the sailors on the boat, seeing the signal, wind this rope in, while at the same time the diver climbs up the mooring-cable (as fast as he can).

From this it would seem that windlasses were employed, and that the waist-rope probably remained attached to the main cable by a loose ring, so that the diver was rapidly brought back to his way of escape when the winding-in began. Tshai Thao continues with a graphic account of the agonies of divers who by accident overstepped the narrow limits of safety, and the means taken to revive them, saying that among those who see and admire pearls in ordinary society, very few have any conception of what it cost to get them. The same emphasis, especially concerning the dangers from sharks and other evil beasts of the sea, is found in a long account by Chou Chhü-Fei [Zhou Qu-Fei] some 60 years later. He adds little to the technicalities however, except to say that baskets are also let down on long cords with the divers themselves, a further precaution since these could be wound in at leisure. One begins to get a picture of slow but continuous improvements in diving technique through the centuries, leading to the inventions of the Ming with which we started.

There is nothing improbable in the techniques described in *The Exploitation of the Works of Nature* (1637). Indeed, they may be quite ancient. A passage from the *Pao Phu Tzu* [*Bao Pu Zi*] (Book of the Preservation-of-Solidarity Master) (c. AD 320) includes, among magical recipes, the following: 'Take a real rhinoceros horn more than 30 centimetres long and carve on it the shape of a fish, then put one end in the mouth and enter the water – the water will open out 90 centimetres on all sides, and you will be able to breathe in the water'. Perhaps this is a concealed reference, in the alchemical manner, to a diver's tube. In any case both breathing-tubes and diving-bells of a kind are alluded to by Aristotle and other ancient writers. A German ballad of 1190 (about Li Chao-Thao's [Li Zhao-Tao's] time) mentions the breathing tube of a diver, and the first European illustration occurs in the work of an anonymous Hussite engineer of about 1430. It is of much interest that between this time and that of Sung Ying-Hsing [Song Ying-Xing], Leonardo da Vinci sketched, in the *Codex Atlanticus*, a breathing-tube such as was used by Indian Ocean pearl-divers; and this, besides having spikes to keep off fishes, is strengthened against collapse under pressure by just such metal rings as referred to in *The Exploitation of the Works of Nature*. But the pressure of the water on the diver's lungs must always have been the great limiting factor for the attempt to use atmospheric air. It is therefore interesting that bellows for pumping it down the tube are mentioned in an Arabic work on hydraulic engineering of about AD 1000, and it would be interesting to know whether the Chinese of the Sung [Song] period used them. The first European mention of a double pipe for breathing in and out occurs in the works of Giovanni Borelli (1679) and Edmond Halley in 1716 combined such pipes with a diving-bell. We have not so far found any references to the latter in Chinese literature, but the idea was current in Europe much earlier than is generally supposed, and remarkable pictures of 'diving-bells', some with breathing-tubes, are to be found in fourteenth and fifteenth century manuscripts of the Alexander-Romance. The adventurous king was supposed to have descended into the depths of the sea in a sphere or barrel of glass, emerging safe and sound in spite of the treachery of his queen Roxana, who let go the cables. The pearling-fields of Asia, that continent which Alexander aspired to rule, especially those of India and China, may well have been the original home of all diving techniques, and their early history may have to be discovered from there. Indeed, it has been plausibly suggested that the breathing exercises so prominently associated with Hindu, Buddhist and Taoist [Daoist] meditational and other disciplines, had a close connection with the practices of the divers who earned a hard living, from antiquity onwards, by ravishing the treasures of an element not natural to man.

So far we have been thinking only of those natural pearls which were sought at such risk by the Cantonese divers century after century. But there are also cultivated pearls and artificial pearls, the former by an intentional insertion of a foreign object inside the mollusc's shell, the latter made entirely by man. The invention of cultured pearls seems to be essentially Chinese. In 1825 J.E. Gray, while studying mollusc shells in the British Museum (Natural History), noticed that some good pearls were still attached to the shell of *Barbala plicata*, and that they had clearly been artificially induced by the introduction of small pieces of nacre ('mother-of-pearl') as the foreign body. They had come from China. Later Gray reported that other examples had been formed round minute pieces of silver wire. Thirty years later there was an eye-witness account of the industry at Huchow [Huzhuo], where foreign bodies of all kinds were used in fresh-water mussels including very small Buddhist images. The local people attributed the invention to a thirteenth-century inhabitant named Yeh Jen-Yang [Ye Ren -Yang].

It is indeed possible to find a clear account in Chinese literature at least a couple of centuries before the time ascribed to this worthy. It occurs in the

Wên Chhang Tsa Lu [*Wen Chang Za Lu*] (Things Seen and Heard by an Official at Court), written by Phang Yuan-Ying [Pang Yuan-Ying] in 1086. He says:

> Hsieh Kung-Yen [Xie Gong-Yan], an Executive Official of the Ministry of Rites, found out a way of cultivating pearls. The way this is done now is to make (first) 'false pearls' (from pieces of nacre, etc.). The smoothest, roundest and most lustrous of these are then selected, and inoculated into fairly large oysters kept in clean sea-water, as soon as they open their valves. The clean sea-water is repeatedly renewed, and at night the oysters take up the best influences of the moon. Then after two years, real pearls are fully formed.

We cannot be sure that Hsieh Kung-Yen was really the first to accomplish this, for certain passages from still earlier books form a kind of background to his achievement. Liu Hsün [Liu Xun] in his *Ling Piao Lu I* [*Ling Biao Lu Yi*] (Strange Southern Ways of Men and Things), finished about 895, gave a brief account of the Lienchow [Lianzhou] pearl fisheries, and makes it clear that in the ninth century the growth of the pearl within the oyster was well-enough understood to suggest inoculation. Indeed, this may have already been practised in some restricted circles, for still earlier we read of presentations of pearls of special shapes. For example, the *Nan Chhi Shu* [*Nan Qi Shu*] (History of the southern Chhi [Qi] Dynasty) records for AD 489: 'The Yüeh [Yue] Prefecture presented (to the throne) a white pearl shaped naturally like the image of a "meditating Buddha", 7.5 centimetres in height.' Even if the size is exaggerated in the record, the nature of the object recalls the practises of the Huchow [Huzhou] people in modern times. Finally, as far back as the second century BC, we have a remarkably perspicacious statement in the *Huai Nan Tzu* [*Huai Nan Zi*] (The Book of (The Prince of) Huai Nan): 'Though luminous pearls are an advantage to us, they are a disease for the oyster'. Once this was realised, the inoculation of an irritant to cause the disease may not have been a far step in thought, though many long years may have passed before anyone managed to do it in practice.

Information about the art spread somehow to Europe before the middle of the eighteenth century, for the great botanist Linnaeus took advantage of it and acknowledged the origin of it. When a young man he saw fresh-water mussel pearling going on in Lapland. Twenty years later (1751) he wrote that he had read of a Chinese method of producing cultured or induced pearls, and in another ten years he had been able to demonstrate the feasibility of the method on the fresh-water mussels of his native Sweden, using small pieces of silver wire together with tiny balls of plaster or limestone. Eventually he sold the process for a substantial sum. The invention was further re-discovered and perfected in Japan where it has become a very great industry.

The real secret of the technique had probably always been to implant small pieces of the nacre-secreting tissue along with the nucleus for causing irritation so that a closed cyst is formed within the mollusc. Naturally this was not revealed to Phang Yuan-Ying [Pang Yuan-Ying]. Otherwise only 'blister-pearls' are formed, as the mollusc's tissue only covers the intruded object inside the shell with mother-of-pearl; most of the Buddhist images are of this kind.

Lastly, to pursue this digression to the furthest end, we ought to consider for a moment the production of purely artificial or imitation 'pearls', not made within a bivalve at all. There is considerable reason for thinking that this technique, like that of the cultivation or induction of true pearls, goes back a long way in China. It involved isolating minute natural pearly crystals and then depositing them as a stable film on spheres made of glass or other material. In Europe it has been current since 1680 when Jacquin, a Parisian rosary-maker, made a preparation called (curiously enough) 'essence d'orient' from the silvery scales of the bleak-fish, *Alburnus lucidus*, a variety of bony carp. Having caused a film of the thick mass of suspended crystals to adhere to the inner walls of small glass drops, he filled the cavity with white wax and thus produced artificial pearls. Now for a century past it has been known that the crystals are those of the purine base guanine, and to this day its lustrous particles are used for the same purpose. But Jacquin's performance strikes a note of memory. Did we not read something of the same kind in connection with ancient Chinese glass? From Wang Chhung's [Wang Chong's] *Lun Hêng* [*Lun Heng*] (Discourses Weighed in the Balance) of AD 83 we did indeed quote the following passage (volume 2 of this abridgement, p. 359):

> Similarly pearls from fishy oysters are like the bluish jade of the Tribute of Yü [Yu]; all true and genuine (natural products). But by following proper timing (i.e. when to begin heating and how long to go on) pearls can be made from chemicals, just as brilliant as genuine ones. This is the climax of Taoist [Daoist] learning and a triumph of their skill . . . .

The whole passage, it will be remembered, concerns the making of mirrors and lenses of glass, imitating not only jade but those highly polished mirrors of bronze which were used to ignite tinder with 'fire from the sun'. It may well be, then, that in ancient times the Taoists also found a way of extracting and suspending the guanine crystals from fish skin, depositing them on glass to make 'false pearls'.

Although the modern sources are primarily the teleostean or bony fishes, the existence of a long-enduring corpus of Chinese legend associating pearls with sharks is just worth noticing. Thus the *Chiao-Chou Chi* [*Jiao-zhou Ji*] (Record of the Chiaochow [Jiaozhou] (District)), a Chin [Jin] work, says that the shark has on its back armour 'patterned with pearls'. The *Phi Ya* [*Pi*

Ya] (New Edifications on (i.e. Additions to) the *Literary Expositor*) of 1096 says that while oyster pearls are in the belly, shark pearls are in the skin. An abundance of texts describe the shark people, who live at the bottom of the sea, give lodging to pearl-divers, and sometimes come ashore to wander about and sell their soft unbleached (pongee) silk. On departing they pay their bills by weeping tears which turn to pearls. Perhaps this is just what certain fishes, in the hands of Taoist [Daoist] alchemists, metaphorically did.

NAVAL TECHNIQUES

The ram

There remains still something to be said about naval techniques as opposed to the peaceful practices of the sea. On various occasions so far we have touched upon the question of whether or not Chinese ships of war in ancient and medieval times engaged in the typical Graeco-Roman technique of ramming. Fundamentally this would seem unlikely on account of the basic blunt-ended and flat-bottomed build, where no keel invited elongation into a sharp under-water weapon of offence. Nevertheless the attachment of one or two such pointed protuberances to hole the enemy under the water-line was perfectly possible, and indeed there is a certain amount of evidence that these were in fact used. It does not appear, however, that this technique ever held the dominating position that it had in the ancient Mediterranean.

Among the ancient classes of warships there may have been one called 'Stomach-strikers' (Thu-wei [Tu-wei]). It appears for instance in a passage from the *Yüeh Chüeh Shu* [*Yue Jue Shu*] (Lost Records of the State of Yue) of about AD 52, now preserved only in collections such as the *Yuan Chien Lei Han* [*Yuan Jian Lei Han*] (Mirror of the Infinite; a Classified Treasure-Chest) encyclopaedia (1710):

> Ho Lu [He Lu] (king of Wu, r. 514 to 496 BC) had an interview with (Wu) Tzu-Hsü [Zi-Xu] and asked him about naval preparedness. (Tzu-Hsü) answered: 'The (classes of) ships are named Tai-i [Dai-yi] (Great wing), Hsiao-i [Xiao-yi] (Little wing), Thu-wei [Tu-wei] (Stomach-striker), Lou chhuan [Lou chuan] (Castled ship) and Chhiao chhuan [Qiao chuan] (Bridge ship). Nowadays in training naval forces we use the tactics of land forces for the best effect. Thus Great-wing ships correspond to the army's heavy chariots, Little-wing ships to light chariots, Stomach-strikers to battering-rams, Castled ships to mobile assault towers and Bridge ships to the light cavalry.'

With one of these types we are already familiar, the castled ships full of marines, hard though it is to say whether oars or sails were their principal means of propulsion. The two 'wing' classes must obviously be sailing ships, and the 'bridge' ships may well have been rowed boats pressed into service from normal use in pontoon bridges. As for the 'stomach-strikers' or 'colliding swoopers' it is difficult to interpret them as anything else than ships fitted with rams. But the date of the passage may not be quite what it seems. Another excerpt purporting to be from the same book occurs elsewhere in the *Thai-Phing Yü Lan* [*Tai-Ping Yu Lan*] (Tai-Ping reign-period Imperial Encyclopaedia (lit. the Emperor's Daily Readings)) of 983; it describes the 'Great-wing' man-o'-war, following what may have been a lost book called *Wu Tzu-Hsü Shui Chan Fa* [*Wu Zi-Xu Shui Zhan Fa*] (Wu Zi-Xu's Manual of Naval Tactics). Each ship, 36 metres long and 4.9 metres in the beam, carried 26 marines and 50 rowers, and various other crew, and an equipment of 4 long 'grappling-irons', 4 spears and 4 long-handled axes, under the command of 4 petty officers, and so on. Similar details occur in another lost book, this time of the twelfth century, where it is remarked that the ships of Chou [Zhou] times must have been quite large. We may feel some hesitation in accepting such a conclusion for the sixth century BC, and prefer to regard these accounts as revealing the practice of the Han San Kuo [Han San Guo] or Chin [Jin] periods, but in any case the presence of rams on certain craft seems fairly sure. At least equally important to note here, however, is the mention of 'grappling-hooks', a subject shortly to lead us in unexpected directions.

No doubt there is an ancient tradition that the fleets of Wu and Yüeh [Yue] had engaged in ramming. From a book of which little is now left, *Wan Chi Lun* [*Wan Ji Lun*] (The Myriad Stratagems) written by Chiang Chi [Jiang Ji] about AD 220 the following words have come down to us:

> When Wu and Yüeh were fighting on the Five Lakes (modern Thai-hu [Taihu]), they used ships with oars, which butted into each other as if with horns. Whether handled bravely or timidly all were overturned, whether blunt or sharp all capsized (and sank).

Rams again, perhaps, yet the description does not quite suggest holing below the water-line. More reliably matter-of-fact is the description in the *Hou Han Shu* (History of the Later Han Dynasty) of the battles on the Yangtze [Yangzi] River in AD 33. In one of these a floating bridge with fortified posts on it, and protected by a boom which blocked the river and by forts on the surrounding slopes, was at last taken by a fleet of several thousand vessels which included castled ships, rowed assault boats and 'colliding swoopers' (Mao-thu [Mao-tu]), and the commentary of 676 explains that the advantage of the latter was that they could butt in violent collision, i.e. that they could ram.

At this point we come to a curious gradual shift in technical terminology. It is clear that the Mao-thu [Mao-tu] ships were the predecessors of the attack ships later called Mêng-chhung [Meng-chong]. Already in AD 100 the *Shih Ming* [*Shi Ming*] (Explanation of Names) dictionary says: '(Vessels that are) long and narrow in appearance are called Mêng-chhung; they dash (like battering-rams) against the ships of the enemy'. As we shall shortly see, by 759 the term Mêng in this combination had come to mean primarily 'armoured' protection for the crew (whether of wet leather, wooden planks or iron plates) and not an impetuous course leading to collision; we shall there translate the two words as 'Covered swooper'. The explanation is that the old word *mao* had two quite distinct meanings, not only 'rushing and colliding' but also 'hat or covering'. By 208, in naval fights between the Wu and Wei forces, Mêng-chhung ships had become common, but it is permissible to think that while the emphasis changed in these centuries from the ram function to the armour build there was a continuity of nomenclature because shipwrights were under the impression that *mao* had always been used in the sense of 'hat' rather then the sense of 'rush'. Presently we shall see how this would fit in with a general trend from close-quarters to projectile warfare discernible all through the history of Chinese sea tactics.

The question has been raised whether another argument for rams in early times would not arise from the term *ko chhuan* [*ge chuan*] (dagger-axe ships, or halberd ships). We have met the term in Han times quite often, taking it in its most obvious meaning as ships manned with marines using dagger-axe halberds. But commentators in the third and seventh centuries have disagreed, some claiming that halberds had been fixed to the hulls of boats. Far be it from us to judge, but may not some memory of ancient ramming tactics be the background of the argument?

That the Chinese build of hull did not lend itself to the sticking on of halberds or any other sharp-pointed part under the water-line is, as has been said, obvious, but in this connection we may recall one most unusual type of boat still extant in modern China which possesses a stem and stern cleft into two parts. This is a sampan used near Hangchow [Hangzhuo], and its origin is quite unknown. With its ram-like projection it resembles the boats of ancient Scandinavia and traditional Indonesia–Polynesia. Perhaps this sampan is a solitary relic of the Indonesian component in Southern Chinese culture, and perhaps also it may be relevant to what evidence there is for the use of the ram in Chinese antiquity.

The passage which has generally been thought to give the clearest proof of rams in the Warring States period seems to us rather to prove something else, and to lead off in a different direction. It occurs in the *Mo Tzu* [*Mo Zi*] (The Book of Master Mo) of the fourth century BC, and tells how the famous engineer Kungshu Phan [Gongshu Pan] came south about 445 BC and re-organised the navy of Chhu [Chu] in its struggle against Yüeh [Yue]:

Master Kungshu then came south from Lu to Chhu and began making naval warfare implements called 'hook-fenders' (*kou chhiang* [gou qiang]). When an (enemy ship) was about to retreat, one used the hook (part); when an (enemy ship) came on, one used the fender (part). The length of this weapon was adopted as a standard for the ships, so that the vessels of Chhu were all standardised while those of Yüeh were not. With this advantage the Chhu people greatly defeated those of Yüeh.

Master Kungshu was proud of his ingenuity and asked Master Mo, saying: 'My warships have the "hook-fender" device (one part to pull and one part to push). Do you have anything like this in your (philosophy of) righteousness?' Master Mo replied: 'The grappling-and-ramming device in my (philosophy of) righteousness is much better than your war-boat gear . . . .'

And Mo Tzu goes off into a sermon (excellent in its way) on 'pulling with love and pushing with respect', for which the whole story was no doubt a pretext. That does not mean of course that we should reject it as a valid account of something that happened in fourth-century BC naval warfare. But the 'hook-fender' device was not exactly a ram; we believe it was a heavy T-shaped iron piece (like a dagger-axe in shape) fitted at the end of a long spar pivoted in derrick fashion at the base of the mast, and capable of being either dropped heavily on the retreating enemy's deck to pin it at a desired distance, or lowered into position to fend off his closing ship at about the same distance away. In both situations the enemy was held at the best crossbow range. This leads directly to our next subject.

Armour-plating and 'grappling-irons'; projectile tactics versus close combat

If, as we believe, the application of thin sheets of metal to the under-water parts of hulls with a preservative purpose was at least as old in China as it was in the Mediterranean region, and if, as we shall hope to show, the military and naval tactics most favoured by the Chinese in all periods tended to be those of projectile exchange rather than those of close hand-to-hand combat, it would be expected that a development of 'armour-plating' above the water-line would occur among them at quite an early date, extending the idea of city walls to the bulwarks of fighting ships. And indeed this is just what we find. In these late medieval contexts, of course, ship-armour does not mean anything like what it came to mean in the world of the nineteenth century, but its perfectly legitimate ancestor may be observed in the thin plates of forged wrought iron which were sometimes fixed to the upper works of Chinese war-junks from the Sung [Song] onwards. Since projectile weapons generally evoke more important works of defence than those required against shock weapons, this development was entirely natural. It was accompanied by another, even stranger for Western preconceptions, namely the use of 'grappling-irons'

Fig. 245. Reconstruction of one of the armoured 'Turtle ships' used by the Korean naval forces under the admiral Yi Sunsin against the Japanese in the last decade of the sixteenth century. The model is in the National Historical Museum at Peking [Beijing]. It is believed that at least two masts were used, but lowered before battle through a central fore-and-aft slot in the armoured roof (not shown in this model).

not for the purpose of laying gangways for boarding-parties, but rather as great pick-axes for holding the enemy ship, or alternatively as clamps for the opposite purpose of keeping her crew at a distance where they could be shot down by point-blank fire.

At several points earlier mention has already been made of the wooden bulwarks which concealed the rowers and sailors on Chinese war-junks, and discouraged boarders. The next logical step was to roof over the whole deck, leaving a slot through which the mast or masts could be raised or lowered, and then to give further protection by armouring the roof and side with iron or copper plates. This was seen fully developed in the fleets of the great Korean admiral Yi Sunsin at the time of the Japanese expeditions of Hideyoshi to Korea (AD 1592 to 1598). For Yi Sunsin then built a number of 'Turtle ships', which proved very effective in the sea battles against the invaders off Chemulpo and Fusan Sound.

It is rather difficult to get a clear idea of these Turtle ships, but though the sources are not very informative, careful study has shown that a typical warship of this kind appears to have been about 33 metres in length with a beam of some 8.5 metres and one main deck 2 metres above the ship's bottom (Fig. 245). The rowers' positions were within the hull, on each side of a central line of cabins for stores and bunks, leaving the main deck free for the gunners and musketeers to maintain fire through 12 gunports and 22 loopholes. They in turn were protected by a sloping roof, the slot in which was probably closed by a sliding hatch before entering battle. The roof was certainly studded with spikes and knives, and though no contemporary text has been found which proves that it was always covered with metal plates, strong local traditions, dating back to the early seventeenth century affirm this. It would at any rate tally with the fact that none of these ships could be set on fire by the incendiary weapons of the Japanese, and that it was difficult, if not impossible, to pierce their walls with the projectiles of the time. On the other hand, it is certain that each Turtle ship carried an animal figure-head with a tube through which dense toxic smoke could be emitted, the result of activities of chemical technicians hidden in the bows. Such compositions, popularly supposed to contain sulphur and saltpetre, had long been known in China.

Perhaps the most interesting feature of the whole episode is that Yi Sunsin was carrying to its ultimate conclusion the projectile tradition of Chinese sea warfare as opposed to shock tactics. The men of his Turtle ships were fully protected against arrows, musket-shots, and incendiary weapons, but also above all against the boarding-parties beloved of the Japanese. Keeping the beam fairly narrow, he built for speed so as to command range, and adding smoke-screen equipment, he gained valuable superiority in surprise manoeuvre. Finally it seems that his armament outweighed the Japanese guns and hand-guns by a factor of forty to one. The tactics were therefore no longer to close and board, but to stand off and pound with projectiles. A line-ahead formation was used to deliver successive broadsides, and ramming might follow only if the enemy was disabled.

About the same time similar factors were leading to similar developments in Europe. During the seige of Antwerp in 1585, the Dutch gave partial protection with iron plates to a man-of-war, but it was much less successful than the Turtle ships of Yi Sunsin, for it went aground at once and was captured by the Spaniards, who made no use of it. Yet in spite of gunpowder, ship-armour was curiously slow in acquiring importance in Europe.

As late as 1796 there was still at least one Korean Turtle ship, with two smoke-emitting heads, at the port of Yohsu in Chulla province. But it is possible to trace this type of vessel, and the tactics which it implies, much further back than Yi Sunsin's time, and hence much further back than the parallel developments in Europe. To begin with the Korean ruler in 1414 inspected a type of warship then thought new, called a Turtle boat. The usual seventeenth and eighteenth-century Chinese texts all contain collections of illustrations of war-junks with varying degrees of protection for crews and marines, but so also does the *Wu Ching Tsung Yao* [*Wu Jing Zong Yao*] (Collection of the Most Important Military Techniques [compiled by Imperial Order]) of 1044. In fact, the preference for projectile tactics against close combat in naval fighting (and the associated appetite for defensive armour) probably goes back as far as one needs to look in China, as far back indeed as the archers of the 'Tower-ships' of the Chhin [Qin] and Han.

The oldest text we have found which seems to have been intended to accompany such a collection of illustrations occurs in the *Thai Pai Yin Ching* [*Tai Bai Yin Jing*] (Canon of the White and Gloomy Planet of War), compiled by Li Chhüan [Li Quan] in 759. It seems worthwhile to give translations of his brief descriptions, for they show rather clearly that as early as the eighth century there was a tendency in Chinese naval construction to roof over completely the upper decks of certain types of combat craft, thereby giving protection from boarders while enabling all projectile weapons to have full play. Here then is what he says:

> *Tower-ships*; these ships have three decks equipped with bulwarks for the fighting-lines, and flags and pennants flying from the masts. There are ports and openings for crossbows and lances, [and at the sides there is provided felt and leather to protect against fire], while (on the topmost deck) there are trebuchets for hurling stones, set up (in appropriate places). And there are also (arrangements for making) molten iron (for throwing in containers from these catapults). (The whole broadside) gives the appearance of a city wall. In the Chin [Jin] period the Prancing-Dragon Admiral, Wang Chün [Wang Jun], invading Wu, built a great ship 200 paces (300 metres) in length, and on it set flying rafters and hanging galleries on which chariots and horses could go. ... ...
>
> *Covered swoopers* (Mêng chhung [Meng chong]); these are ships which have their backs roofed over and (armoured with) a covering of rhinoceros hide. ... ... also both fore and aft, as well as to port and starboard, there are openings for crossbows and holes for spears. Enemy parties cannot board (these ships), nor can arrows or stones injure them. This arrangement is not adopted for large vessels because higher speed and mobility are preferable, in order to be able to swoop suddenly on the unprepared enemy. Thus these (Covered swoopers) are not fighting-ships (in the ordinary sense).
>
> *Combat-junks*; combat-junks have ramparts and half-ramparts above the side of the hull, with the oar ports below. One and a half metres from the edge of the deck (to port and starboard) there is set a deckhouse with ramparts above it as well. This doubles the space available for fighting. There is no cover or roof over the top (of the ship). ... ...
>
> *Flying barques*; another kind of fighting ship. They have a double

row of ramparts on the deck, and carry more sailors (lit. rowers) and fewer soldiers, but the latter are selected from the best and bravest. These ships rush back and forth (over the waves) as if flying, and can attack the enemy unawares. They are most useful for emergencies and urgent duty.

Sea-hawks (or *Sea-grebes*); these ships have low bows and high sterns, the forward parts (of the hull) being small and the after parts large, like the shape of the *hu* bird (when floating on the water). Below deck level, both to port and starboard, there are 'floating boards' shaped like the wings of the *hu* bird. These help the (Sea-hawk) ships, so that even when wind and wave arise in fury, they are neither (driven) sideways, nor overturn. Covering over and protecting the upper parts on both sides of the ship are stretched raw ox-hides, as if on a city wall. . . .

Li Chhüan then goes on to describe 'patrol boats' for collecting intelligence. These also have ramparts.

This text seems to show us that the general principle of projectile warfare at sea from 'armoured' ships which could approach their targets rapidly, deliver a broadside, and make away again, can be traced back to the eighth century at least. Although the Covered swoopers, surely the lineal ancestors of the Turtle ships of Yi Sunsin, are said to be hardly fighting ships in the strict sense, while the Combat-junks are specifically so termed, this may have been an explanation intended for military leaders used to close combat on *terra firma*, and it would be unwise to conclude that grappling and boarding had always been typical of Chinese naval practice in earlier times. Indeed the term Mêng-chhung [Meng-chong] itself as a designation of naval vessels goes back at least to the second century AD. Then the tendency to cover over the upper deck appears also in the fifth type, the Sea-hawks, which from the description suggest some kind of converted cargo-boat like the river-junks in the Sung [Song] pictures. And the great majority of the crew and soldiers, except perhaps the artillerists on the topmost deck, were clearly protected in the battleships known as 'Tower-ships'.

So much for the general principle of ship-armour. But there is more than this for we have several Chinese records of plating with iron considerably earlier than the time of Yi Sunsin. One scene was at the end of the Yuan period. In 1370 a Western Expediton was mounted against Ming Shên [Ming Shen] and began to make its way up the Yangtze [Yangzi], and the *Ming Shih* [*Ming Shi*] (History of the Ming Dynasty) says:

Next year Liao Yung-Chung [Liao Yong-Zhong] was deputy commander of the Western Expedition, ... (he) arrived at Old Khueifu [Kuifu] and routed its defenders under the Szechuanese

[Sichuanese] general Tsou Hsing [Zou Xing] and others; then going on he reached the Chhü-thang [Qutang] Gorge. Here, where the cliffs are very precipitous and the water most dangerous, the Szechuanese had set up iron chains (as booms), and bridges, to block the gorge horizontally so that no ships could get through. (Liao) Yung-Chung therefore secretly sent several hundred men with supplies of food and water to make a portage with small boats, so that they appeared up river beyond these defences. Now the mountains of Szechuan [Sizhuan] are so well wooded that he had ordered the soldiers to wear green garments and sleeveless raincloaks made of leaves; and thus they descended through the forests and rocks. When agreed stations had been reached, the best troops were ordered to attack at Mo-yeh [Mo-ye] Ferry. At the fifth night watch the general assault began both by water and land. The bows of the naval ships were sheathed in iron and all kinds of firearms were made ready on them. Only as dawn was breaking was the army's presence discovered by the Szechuanese, who threw in all their best troops for the defence, but unavailingly.

This engagement was certainly a remarkable one, and much credit must go to the defeated Szechuanese for the ingenuity and engineering skill of their defences. But the use of iron ship armour by Liao Yung-Chung is also of interest since it occurred a couple of centuries before the Turtle ships of Yi Sunsin.

It was however very far from new in 1370. For at the time of most rapid development of the Southern Sung navy, in 1203, a remarkable shipwright, Chhin Shih-Fu [Qin Shi-Fu], built at the Chhihchow [Chizhou] yards two prototype 'Sea-hawk' paddle-wheel warships the sides of which (and perhaps the flat roofs also) were armoured with iron plates. The decks were given complete protection by the overhead cover, and besides the usual arrangements for crossbows, fire-lances, bomb-throwing trebuchets, etc., each ship was provided with a spade-shaped ram. The smaller vessel, of 100 tonnes burden, and with two treadmill-operated paddle-wheels, needed a propulsion crew of 28 men; the larger of some 250 tonnes though not very much greater in length, and apparently with four paddle-wheels, needed 42 men to work them and carried 108 marines. The paddle-wheels were protected by housing above the water-line.

To defend the propulsion mechanism as well as the projectile-firing combat elements themselves was a very natural and logical step in the development of armour, but (failing the invention of the screw-propeller) paddle-wheels alone permitted it. It could not be done either for masts or for projecting oars. There is thus more connection than might at first sight appear between the relatively early flourishing both of paddle-wheel propulsion and of ship-armour in Chinese culture. The history of the paddle-wheel boat will be discussed in a later volume of this abridgement, where it will be seen that this motive power suited the tactics of rapid fire from a fast armoured ship. When paddle-wheel steam warships first approached the coast of China in the early nineteenth century they brought nothing tactically novel, indeed they constituted the realisation of a dream which projectile-minded Chinese admirals had entertained for well over a millennium. Although the cruisers of Chhin Shih-Fu [Qin Shi-Fu] were said to be of 'new design', there is no special reason for regarding them as the first of all Chinese iron-armoured ships, and earlier examples may well come to light, though – at a guess – perhaps not before the twelfth century AD, when the foundation of a permanent navy occurred.

We can follow out the contrast between projectile tactics and close-combat boarding-party tactics much further still in the context of 'grappling-irons'. If we trace down the warship descriptions of the eighth century AD in later compilations of naval knowledge, we find that the *Wu Ching Tsung Yao* [*Wu Jing Zong Yao*] (Collection of the most important Military Techniques [compiled by Imperial Order]) of 1044 inserted some very interesting material from a quite independent, and still older, source. Under its entry for Patrol boats it included a description of the Wu-ya hsien [Wu-ya xian] (Five-Ensign battleships) which should more properly have been put with the Castled battleships. After a slightly abridged version of the Thang [Tang] text on Patrol boats, the passage continues by quoting from the *Sui Shu* (History of the Sui Dynasty), the biographical parts of which were finished by AD 636. It tells the story of the building of fleets for the emperor Kao Tsu [Gao Zu] by a celebrated engineer Yang Su, and of the naval battle in which they sealed the fate of the Chhen [Chen] dynasty:

> [In the 4th year of the Khai-Huang [Kai-Huang] reign period (584 AD)] Sui Kao Tsu commissioned Yang Su (as commander-in-chief) to destroy the Chhen. So coming down the gorges to Hsinchow [Xinzhou] he built [at Yung-an [Yong-an]] great war-junks called Wu-ya hsien (Five-Ensign battleships). Above they had five decks, and their height was more than 30 metres. To left and right, and fore and aft, he set up six *pho-kan* [*po-gan*] (lit. striking- or patting-arms), each 15 metres long. And of marines (each ship carried) 800 men, and many flags and banners fluttered aloft.
>
> ... When the ship comes alongside an enemy ship, they let go the striking-arms on top of her, and whether barque or barge she is all broken into fragments.
>
> ... he ordered the Men of Pa (highlanders in Szechuan [Sichuan]) to man four of the Five-Ensign battleships and fight them with the striking-arms. They thus destroyed more than ten of the great war-

junks of the Chhen fleet, breaking them to pieces, and so the river route was forced open.

What were these 'striking-arms?' It seems fairly certain that they were long heavy pointed spikes, probably iron-shod, and fixed at right angles at the end of elongated arms like guy-derrick jibs, then suddenly released from an approximately vertical position to crash down through the deck and hull timbers of the enemy vessel. They were certainly not intended to 'grapple', i.e. to provide gang-planks for boarders, and the best name for them would be 'holing-irons'. We can actually see them, rather badly drawn as usual, in Fig. 246, looking like thin hammers, and in the unarmed position.

If the shafts of the holing irons were long enough, and if the enemy crews were not provided with gangways of the right length known beforehand, then even if they wanted to attack by boarding, they would be held, as it were, at arm's length, where the missiles of the battleship could play upon them and decimate them. Such 'grappling-irons for killing people off', or what we may call 'fending-irons', are mentioned by Lu Yu [Lu You] about 1190 in his *Lao Hsüeh An Pi Chi* [*Lao Xue An Bi Ji*] (Notes from the Hall of Learned Old Age) where he describes the campaign government forces had to wage some 60 years earlier against an important equalitarian revolt. 'The rebels', he says, 'had fighting-ships . . .' and he goes on:

> Among their offensive weapons were the *na tzu* [*na zi*] [commonly pronounced *nao tzu* [*nao zi*]], the 'fish forks' (*yü chha* [*yu cha*]), and the *mu yao la*. The *nao tzu* and the *yü chha* were on bamboo poles 6 to 9 metres long like handles, and prevented the (government) marines with hand-weapons from boarding and attacking at close quarters. . . . The *mu lao ya*, which was also called the 'no *corvée* log' (*pu chieh mu* [*bu jie mu*]), was a piece of wood strong and heavy just a little over 0.9 metres long and sharpened at both ends; these were used by the warships (on both sides?) and proved very effective . . . .

The *na tzu*, 'kneaders' or 'pounders', were probably smaller versions of the holing-irons, and succeeded in keeping the enemy at an inconvenient distance. Another writer of the period tells us that the rebel ships mounted what might be called 'holing-derricks'. In his *Chi Yang Yao Pên Mo* [*Ji Yang Yao Ben Mo*] (The History of (the Rebellion of) Yang Yao from Beginning to End), written about 1140, Li Kuei-Nien [Li Gui-Nian] says:

> The rebel ships had two or three decks, . . . They were equipped with 'holing-derricks' (*pho-kan* [*po-gan*]), which were like great masts over 30 metres high. Large rocks were hoisted up to the top of these by means of pulleys, and when a government ship came close, they were suddenly let go to smash her . . .

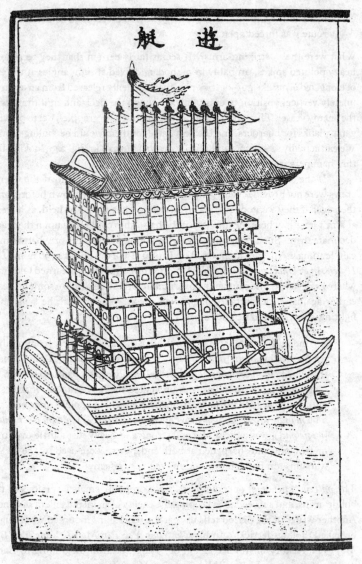

遊 艇

Fig. 246. An illustration of the 'Five-Ensign' battleship, the Wu-ya hsien [Wu-ya xian]. From the Chhing [Qing] edition of the *Collection of the most important Military Techniques* [*compiled by Imperial Order*] but originally described in the *Sui Shu* (History of the Sui Dynasty) (AD 636). Although the highly fanciful creation of an architectural painter, it is valuable for one thing – it is almost the only traditional drawing we have which shows the 'striking-arms' or 'holing-irons', or 'fending-irons', three on the port side in the lowered position. Though very badly drawn, we know that these were heavy pointed spikes on the ends of long spars.

Here then, though the same technical term is used as for the holing-irons of the Sui, the projectile mentality had again overcome the conception of a shock weapon, and heavy weights were dropped from jib arms above in the attempt to hole and sink the opposing craft.

It is interesting that an exactly similar device is recorded as having been in common use at the same time, and perhaps somewhat earlier, in the Byzantine navy. It had Hellenistic precedents. But what a contrast all these Chinese 'grappling-irons' were with the celebrated *corvus* of the Romans. In the wars with the Carthaginians, *c*. 260 BC, the Roman tacticians decided to accept the necessity of head-on collisions, but to avoid the usual sequel in which the more agile ship would disengage and fatally ram the less agile. This they did by a device which would fix the ships together and at the same time give scope for their own superiority in hand-to-hand fighting by pouring their marines on to the Carthaginian decks. It consisted of a gangway in the bows, 11 metres long and over a metre wide, armed underneath at its outboard end with a massive iron spike, and held in readiness by suspension over a pulley at the top of a 7 metre post. When the plank had fastened itself to the enemy ship, the Roman legionaries could board it two abreast. The comparison of their short swords with the crossbows and catapults of the Chinese gives the measure of the difference between two utterly different conceptions of naval warfare.

A word may be added here about the mounting on Chinese warships of trebuchet or mangonel artillery (i.e. catapults using a long arm pulled down sharply at one end, in the Chinese case, either by manned ropes or a counter-weight). This is clearly attested for in the eighth century in words which we read from the *Manual of the White and Gloomy Planet of War*; (*Venus* (p. 259)), and all the subsequent compendia support it. At that time, and also in the eleventh century when the *Collection of the most important Military Techniques* was compiled, the trebuchets must have been of the classical manned type, but the only illustrations we have are later, and show the counterweighted type (Fig. 189). There can be no doubt that this was a constant practice through the ages.

From the foregoing pages only one character still needs pinning down – the Prancing-Dragon Admiral, Wang Chün [Wang Jun]. What he did in AD 280 prefigured very closely the great naval victory of Yang Su in 585 about which we have just been reading, for in both cases a powerful fleet descending the Yangtze [Yangzi] from the west overbore all the defences both of land and water which had been set up against it, and brought about a change of dynasty. Needless to say, the three hundred years that intervened saw many changes in ship-building, and though in the third century we hear nothing of holing-arms or paddle-wheel boats, we do find other techniques equally interesting, no-tably the building of floating fortresses borne on multiple hulls. This has perplexed many students because the seemingly extraordinary dimensions found their way into later texts describing Castled battleships, as we have seen. But the matter is soon unravelled when we look at the biography of Wang Chün in the *Chin Shu* [*Jin Shu*] (History of the Chin [Jin] Dynasty), completed in 635. It runs as follows:

> The emperor Wu Ti [Wu Di] (of Chin), meditating the overthrow of the Wu State, ordered (Wang) Chün to construct (a fleet of) warships. (Wang) Chün also built a great square composite floating fortress with multiple hulls, 120 paces (183 metres) along each side, holding more than 2000 men, with high towers and four sally-port gates, set in wooden walls round the ramparts on which horses could be ridden back and forth. At the bows there were decorations of strange birds and animals to overawe the river spirits. For abundance of oars and skill of ship-craft, nothing like it had ever been seen before.
>
> (Wang) Chün built his ships and the floating fortress in Szechuan [Sichuan] from the wood of persimmon trees, and some of the (smaller) timbers floated down the (Yangtze) river. The Wu governor of Chien-ping [Jianping] Wu Yen [Wu Yan], noticed this, had the pieces collected from the water, and showed them to Sun Hao (the last emperor of Wu), saying that the Chin ruler was evidently bent on war and advising an urgent increase in the army of defence. But his advice was not taken.

The rest of the story is told in the pages of the *History*. The Chin army and navy did indeed descend from the Yangtze [Yangzi] in force. The Wu defenders used iron-chain booms, and iron stakes more than three metres long stuck in the shallows to hole the attacking ships. But Wang Chün also built several dozen large rafts, each more than 100 paces (152 metres) square, with dummy marines wearing what looked like armour and carrying fake weapons; these rafts piloted by skilled sailors, crashed over the iron spikes and bent or broke them. Then with the aid of quantities of hemp-seed oil poured on the water at several points, and ignited by torches more than 30 metres long, the boats supporting the chain booms were set on fire, so that the chains themselves melted in the heat and no further obstacle remained to hold back the ships and the floating fortress. No doubt this huge craft might also be termed a floating battery, for there is evidence that trebuchet catapults were almost certainly mounted upon it. It is a pity that no details were given of the joint hulls by which it was supported, but we may surmise some 16 vessels each about 46 metres long. The idea could have arisen very naturally from the floating-bridge principle, which was much used in Chinese antiquity. Although these facts have no bearing on the largest size ever reached by Chinese ships, as once was thought, they do supply yet another indication of the prevalent projectile-mindedness – if we have no strong-points or artillery emplacements on the frontiers of the Wu territory, said the Chin strategists, we shall built a great one and float it right down into their midst.

CONCLUSIONS

The most suitable ending for these chapters on nautical technology will perhaps be a brief tabulation of the characteristics of Chinese craft, and the influence which they may have exerted upon Western practice throughout the centuries. Let us begin with the view, so highly probable, that the basic principle of Chinese ship construction was derived from the example of the bamboo stem with its partitions or septa, and indeed that in actual fact the earliest vessels of East Asia were rafts of bamboo. This led directly to (A) the rectangular horizontal plan. The following consequences resulted:

(i) The absence of stem-post, stern-post and keel.
(ii) The presence of bulkheads, giving a hull resistant to deformation, and leading naturally to
(iii) The system of water-tight compartments, with its many advantages. These were almost surely in use by the second century AD, but were not adopted in the West until the end of the eighteenth. Provenance was then recognised.
(iv) The possibility of free-flooding compartments, found useful both on river rapids and at sea. This was not adopted to any extent in Europe.
(v) The existence of a vertical member to which the axial rudder could be attached, in 'line closure' rather than 'point closure'. See (D) (i).

Additional elements not essential to the design were:

(vi) Approximation to a flat bottom. This goes back many centuries in China, but was not adopted in Europe for ships of any size until the nineteenth century.
(vii) Approximation to a rectangular cross-section. This was also old in China, but again not adopted in Europe until the development of ships of iron and steel.
(viii) The placing of the ribs enclosing the greatest area well aft. This is still prevalent in traditional Chinese ships, and must be old, though how old has not yet been established; it was certainly current in the Thang [Tang] (eighth century AD) and may well have been earlier. Its interest for sailing vessels was not understood in the West until the end of the nineteenth century.

With regard to propulsion, Chinese seamanship had the lead of Europe for more than a millennium. First, concerning the use of (B) oars and paddles, we note:

(i) The invention, not later than the first century AD, of the self-feathering 'propeller' or sculling-oar (the yuloh). Though universally used in China, this was never adopted in the West.

(ii) The invention of the treadmill-operated paddle-wheel boat in the eighth if not the fifth century AD, and its great development in the Sung [Song] (twelfth century) for warships with multiple paddle-wheels and catapult artillery. Though proposed in the fourth century AD in Byzantium, and discussed in Western Europe in the fourteenth and fifteenth centuries, no practical use of the principle was made until the sixteenth century in Spain.

(iii) The complete absence of the multi-oared galley from Chinese civilisation, whether powered by slaves or free-men (apart from small patrol craft and the paddled dragon-boats used only for ritual races). This must partly be regarded as the result of a relatively advanced development of

(C) sails and rig. Here several points are important:

(i) From at least the third century AD onwards, ships of the Chinese culture-area were fitted with multiple masts. This may well have been a consequence of (A)(ii) above, since the bulkheads invited the placing of several 'tabernacles' to hold the masts along the fore-and-aft mid-ship line. Europeans of the thirteenth century and later were greatly impressed by the large size and many masts of the sea-going junks, and in the fifteenth century they adopted a system of three, which led in due course to the development of the full-rigged ship.

(ii) The Chinese also staggered their masts across the width of the ship in order to avoid the becalming of one sail by another. This is approved by modern sailing ship designers, but was not adopted by Europeans during the period of importance of the sailing ship. Nor did the Chinese practice of radiating the rakes (tilts) of the masts like the spines of a fan win acceptance in other parts of the world.

(iii) One of the earliest solutions of the problem of sailing to windward in large vessels was due to the Chinese of the second and third centuries AD, or to their immediate Malaysian and Indonesian neighbours at the zone of Sino–Indian culture-contact. This involved the development of fore-and-aft sails. The Chinese lug-sail arose, in all probability, from the Indonesian canted square-sail, and hence indirectly from the square-sail of ancient Egypt; perhaps also, as language suggests, it had something to do with the 'double-mast sprit-sail' (now known only in Melanesia), which in its turn had developed from the Indian Ocean 'twin-mast sprit-sail'. The Chinese sprit-sail would have had similar origins. Roman–Indian contacts at the same period (second and third centuries AD) generated the sprit-sail in Mediterranean waters, but it seems to have fallen out of use there, and was introduced a second time from Asia at the beginning of the fifteenth century. Meanwhile, in the West, the lateen sail, characteristic of the Arab culture-area, dominated in the Mediterranean from about the end of the eighth century, and spread to the ocean-going full-rigged ship in the latter part of the fifteenth. The later lug-sails of Europe derived in all probability from Chinese balance lugs.

(iv) The earliest tightly bent 'aerofoil' sails were the mat-and-batten sails developed in China from the Han period onwards. The system involved many ingenious auxiliary techniques, such as multiple sheeting. Such sails were never used in the West during the period of the importance of the sailing ship, but modern research has demonstrated their value, and present-day racing yachts have adopted important elements of the Chinese rig, including battens for tautening the sails, and the system of multiple sheets.

(v) As a consequence of (A)(vi) above, and having regard to (C)(iii), the use of leeboards and centre-boards developed in China, probably from the sailing-rafts of ancient times, and certainly by the Thang [Tang] (seventh century AD). They were adopted in Europe a thousand years later, in the late sixteenth century. Sailors of the Chinese culture-area raised and lowered them by tackle (especially when the large rudder served the purpose), by pivoting, or by sliding in grooves.

In the domain of vessel control, there was the great development of the steering-oar which centred on the invention (D) of the rudder. Remembering (A)(v) above, we note:

(i) The fundamental invention of the axial or medial rudder. This device was fully developed in China by the end of the second century AD (probably the first century), and the attachment to the stern cross-piece, if not already achieved, followed soon afterwards, certainly by the end of the fourth; but its first appearance in Europe did not occur until the end of the twelfth century. Steering-oars and stern-sweeps, however, though thus relegated to a secondary position so early in Chinese nautical technology, were never completely abandoned, partly because of their value in negotiating river rapids, partly for manoeuvring when the ship was nearly stationary. Bow-sweeps lived on for the same reasons.

(ii) The invention of the balanced rudder, more efficient in the water than the unbalanced type. This was current in China at least as early as the eleventh century AD, but was still regarded as a new and important device at the end of the eighteenth century in Europe.

(iii) The further invention of the fenestrated rudder, also more advantageous in its behaviour in the water. This was not adopted in Europe until the era of iron and steel ships.

Among miscellaneous ancillary techniques (E), the following are worthy of remark:

(i) Hull sheathing. Already in the eleventh century AD the superimposition of layers of fresh strakes was usual in China; in Europe it became general in the sixteenth and later. Sheathing with copper plates was discussed, if not actually performed, in the fourth century AD in China; lead had been used in Hellenistic Europe. The practical use of copper did not become general, both in East and West, until the eighteenth century.

(ii) Armour plating. The strong predilection of Chinese naval commanders throughout the ages for projectile tactics as opposed to reliance on the close combat of boarding-parties led to the armouring of hulls and upper-works above the water-line with iron plates by the end of the twelfth century, and this continued in later times with notable Korean contributions in the sixteenth. By that time similar developments were occurring, though rather half-heartedly, in Europe.

(iii) The development of non-fouling 'stockless' anchors.

(iv) The invention of articulated or coupled trains of vessels, probably in the sixteenth century AD. This was not often employed in Europe for shipping, but greatly used in other fields of transportation.

(v) Ingenious dredging practices.

(vi) Advanced techniques of diving generated by the pearling industry.

(vii) Advanced techniques of bilge clearance by chain-pumps, admired by sixteenth century Europeans.

So much for the genius of Chinese shipwrights and sailors. Of the most ancient influences acting upon their ship construction, we have been able to detect certain affinities with the naval architecture of ancient Egypt. These include (*a*) square-ended hulls, (*b*) two-legged masts, (*c*) the anti-hogging truss in dragon boats (to prevent bow and stern portions from drooping), (*d*) stern galleries, (*e*) some forward-curving sterns, (*f*) the practice of facing rowing. If we knew more about the shipping of ancient Mesopotamia we might find that some of these radiated outward from there in both directions, yet on the whole the people of ancient Egypt seem to have been water-farers cleverer and more assiduous than any nations of the Fertile Crescent – save the Phoenicians, and they were much younger. Of Chinese connections with South-east Asian practices, the most important were (*a*) the interchanges in sails and rig, (*b*) the multi-paddled dragon-boats, and (*c*) the poling gangways, which may be connected closely with outriggers.

During the past two thousand years there seems to have been scarcely any century which did not witness the transmission of one or another element of nautical technology from Asia to Europe. It does no harm for us to realise this, children as we are of that western archipelago where maritime commerce arose and flourished so exceedingly, whence the *conquistadores* set forth on their explorations of every strait and ocean. The succession may be sketched as follows:

Second century AD	The sprit-sail (from India to the Roman Mediterranean).
Eighth century	The lateen sail (from the Arabic culture-area to Byzantium).
Late twelfth century	The mariner's compass and the axial (sternpost) rudder (either by Arabic-Crusader contact or overland through the West Liao State in Sinkiang [Xinkiang]).
Seventh to fifteenth centuries	Preconstructed rib frames as opposed to strake-morticed hulls with inserted frames, possibly derived from bulkhead construction.
Fifteenth century	Multiple masts (from Chinese junks), the sprit-sail again (perhaps from Sinhalese craft), and the adoption of the lateen, first on all masts, then on the mizen with square-sails on the other masts.
Sixteenth century	The protection of the hull by additional strake layers.
Late sixteenth century	Leeboards.
Late eighteenth	Water-tight compartments, centre-boards, perhaps also copper hull-sheathing.
Nineteenth and twentieth centuries	Flat bottoms, hulls of approximately rectangular cross-section, balanced rudders, fenestrated rudders, non-fouling stockless anchors, aerodynamically efficient sails, thwartwise staggered masts, multiple sheets, and the placing of the ribs enclosing the greatest area aft of the midship line.

Some of these developments may of course have been partly independent. Even when we have good reason to believe in transmission, we know very little of the means by which it took place. But as in all other fields of science and technology the onus of proof lies upon those who wish to maintain independent invention, and the longer the period elapsing between the successive appearances of a discovery or invention in the two or more cultures concerned, the heavier that onus generally is. The techniques here discussed were, to be sure, often improved by the Europeans when they later adopted them. All that our analysis indicates is that European seamanship probably owes far more than has generally been supposed to the contributions of the sea-going peoples of East and South East Asia. One would be ill-advised to undervalue the Chinese sea-captain and his crew. Those who know them best today are very willing to apply to them the words of an English sea-poet when he compared past and present on the waters:

> The thranite now and thalamite are pressures low and high,
> And where three hundred blades bit white the twin propellers ply,
> The God that hailed, the keel that sailed, are changed beyond recall,
> But the robust and Brassbound Man, he is not changed at all!

Table of Chinese Dynasties

夏	HSIA [XIA] kingdom (legendary?)		*c.* −2000 to *c.* −1520
商	SHANG (YIN) kingdom		*c.* −1520 to *c.* −1030
周	CHOU [ZHOU] dynasty (Feudal Age)	Early Chou [Zhou] period	*c.* −1030 to −722
		Chhun Chhiu [Chun Qiu] period	−722 to −480
		Warring States (Chan Kuo [Zhan Guo]) period 戰國	−480 to −221
First Unification	秦 CHHIN [QIN] dynasty		−221 to −207
漢 HAN dynasty	Chhien Han [Qian Han] (Earlier or Western)		−202 to +9
	Hsin [Xin] interregnum		+9 to +23
	Hou Han (Later or Eastern)		+25 to +220
三國 SAN KUO [S GUO] (Three Kingdoms period)			+221 to +265
First	蜀 SHU (HAN)	+221 to +264	
Partition	魏 WEI	+220 to +265	
	吳 WU	+222 to +280	
Second	晉 CHIN [JIN] dynasty: Western		+265 to +317
Unification	Eastern		+317 to +420
	劉宋 (Liu) SUNG [SONG] dynasty		+420 to +479
Second	Northern and Southern Dynasties (Nan Pei chhao [Nan Bei chao])		
Partition	齊 CHHI [QI] dynasty		+479 to +502
	梁 LIANG dynasty		+502 to +557
	陳 CHHEN [CHEN] dynasty		+557 to +589
	魏 Northern (Thopa [Touba]) WEI dynasty		+386 to +535
	Western (Thopa) WEI dynasty		+535 to +556
	Eastern (Thopa) WEI dynasty		+534 to +550
	北齊 Northern CHHI [QI] dynasty		+550 to +577
	北周 Northern CHOU [ZHOU] (Hsienpi [Xienbi]) dynasty		+557 to +581
Third	隋 SUI dynasty		+581 to +618
Unification	唐 THANG [TANG] dynasty		+618 to +906
Third	五代 WU TAI [WU DAI] (Five Dynasty period) (Later Liang, Later Thang [Tang] (Turkic), Later Chin [Jin] (Turkic), Later Han (Turkic) and Later Chou [Zhou])		+907 to +960
Partition			
	遼 LIAO (Chhitan [Qidan] Tartar) dynasty		+907 to +1124
	West LIAO dynasty (Qarā-Khiṭāi)		+1124 to +1211
	西夏 Hsi Hsia [Xi Xia] (Tangut Tibetan) state		+986 to +1227
Fourth	宋 Northern SUNG [SONG] dynasty		+960 to +1126
Unification	宋 Southern SUNG [SONG] dynasty		+1127 to +1279
	金 CHIN [JIN] (Jurchen Tartar) dynasty		+1115 to +1234
	元 YUAN (Mongol) dynasty		+1260 to +1368
	明 MING dynasty		+1368 to +1644
	清 CHHING [QING] (Manchu) dynasty		+1644 to +1911
	民國 Republic		+1912

N.B. When no modifying term in brackets is given, the dynasty was purely Chinese. During the Eastern Chin period there were no less than eighteen independent States (Hunnish, Tibetan, Hsienpi, Turkic, etc.) in the north. The term 'Liu chhao' [Liu chao] (Six Dynasties) is often used by historians of literature. It refers to the south and covers the period from the beginning of the 3rd to the end of the 6th centuries AD, including (San Kuo) Wu, Chin, (Liu) Sung, Chhi, Liang and Chhen. The minus sign (−) indicates BC and the plus sign (+) is used for AD.

BIBLIOGRAPHY

Anderson, R.G. and Anderson, R.C., *The Sailing Ship; Six Thousand Years of History*, Harrap, London, 1926.

Anon., *Illustrated Catalogue of the Maze Collection of Chinese Junk Models in the Science Museum, London*, London, 1938.

Anon., 'On Watertight Compartments in Ships', *Mechanics Magazine*, 1824, **2**, 224.

Anon. [perhaps Capt. Kellett], *Description of the Junk 'Keying', printed for the Author and Sold on Board the Junk, Such*, London, 1848.

Andrade, E.N.da C., 'The Early History of the Permanent Magnet', *Endeavour*, 1958, **17**, 22.

Audemard, L., (with the assistance of Shih Chun-Shêng [Shi Zhun-Sheng]), *Les Jonques Chinoises; I. Histoire de la Jonque* (posthumously edited by C. Noteboom), Museum voor Land- en Volken-Kunde & Maritiem Museum, Prins Hendrik, Rotterdam, 1957.

Audemard, L., *Les Jonques Chinoises; II. Construction de la Jonque*. As above, 1959.

Audemard, L., *Les Jonques Chinoises; III. Ornementation et Types*. As above, 1960.

Audemard, L., *Les Jonques Chinoises; IV. Description des Jonques*. As above, 1962.

Audemard, L., *Les Jonques Chinoises; V. Haute Yang-tse Kiang*. As above 1963.

Audemard, L., *Les Jonques Chinoises; VI. Bas Yang-tse Chiang*. As above, 1965.

Beazley, C.R., *Prince Henry the Navigator*, Putnam, New York, 1895; London, 1923.

Bowen, R. le B., 'Arab Dhows of Eastern Arabia', *American Neptune*, 1949, **9**, 87. Also, separately, as pamphlet, enlarged, privately printed, Rehoboth, Mass., U.S.A., 1949.

Bowen, R. le B., 'Eastern Sail Affinities', *American Neptune*, 1953, **13**, 81 and 185.

Bowen, R. le B., 'Boats of the Indus Civilisation', *Mariner's Mirror*, 1956, **42**, 279.

Bowen, R. le B., 'The Origins of Fore-and-Aft Rigs', *American Neptune*, 1959, **19**, 155 and 274.

Bowen, R. le B., 'Early Arab Ships and Rudders', *Mariner's Mirror*, 1963, **49**, 303. (A consideration of the al-Hariri picture.)

Brindley, H.H., 'Primitive Craft; Evolution or Diffusion?', *Mariner's Mirror*, 1932, **18**, 303.

Brindley, H.H., 'The Evolution of the Sailing Ship', *Proceedings of the Royal Philosophical Society of Glasgow*, 1926, **54**, 96.

Brindley, H.H., 'Mediaeval Rudders' [and the earliest sprit-sail rig], *Mariner's Mirror*, 1926, 12, 211, 232, 346; 1927, 13, 85.

Brindley, H.H., 'The "Keying"' [a Chinese junk which sailed round the world in 1848], *Mariner's Mirror*, 1922, 8, 305.

Brindley H. H. 'Chinese Anchors', *Mariner's Mirror*, 1924, 10, 39.

B[romhead], C. [E.] N., 'Alexander Neckham on the Compass Needle, *Geographical Journal*, 1944, 104, 63; *Terrestrial Magnetism and Atmospheric Electricity* (continued as *Journal of Geophysical Research*), 1945, 50, 139.

Brooks, C.W., 'A Report on Japanese Vessels Wrecked in the North Pacific Ocean, from the Earliest Records to the Present Time', *Proceedings of the California Academy of Sciences*, 1876 (1875), 6, 50.

Brown, Lloyd A., *The Story of Maps*, Little Brown, Boston, 1949.

Casson, L., *The Ancient Mariners; Sea-farers and Sea Fighters of the Mediterranean in Ancient Times*, Gollancz, London, 1959.

Chapman, S. and Harradon, H.D., 'Archaeologica Geomagnetica; Some Early Contributions to the History of Geomagnetism: I, The Letter of Petrus Peregrinus de Maricourt to Syergus de Foucaucourt, Soldier, concerning the Magnet (AD 1269)', *Terrestrial Magnetism and Atmospheric Electricity*, 1943, 48, 1, 3.

Chatterton, E.K., *Sailing Ships, the Story of their Development from the Earliest Times to the Present Day*, London, 1909; *Sailing Ships and Their Story*, London, 1923.

Chatterton, E.K., *Fore and Aft; the Story of the Fore-and Aft Rig from the Earliest Times to the Present Day*, Seeley Service, London, 1912; 2nd edition 1927.

Clissold, P., 'Early Ocean-going Craft in the Eastern Pacific', *Mariner's Mirror*, 1959, 45, 234.

Clowes, G. S. Laird, *Sailing Ships; their History and Development as illustrated by the Collection of Ship Models in the Science Museum*, Part 1 Historical Notes, Science Museum, London, 1932 (reprinted 1951). [Part 2 is the Catalogue of Exhibits.]

Clowes, G.S. Laird and Trew, C.G., *The Story of Sail*, Eyre and Spottiswoode, London, 1936.

Collis, M., *The Grand Peregrination; the Life and Adventures of Fernão Mendes Pinto*, Faber, London, 1949.

Cordier, H., 'L'Extrême Orient dans l'Atlas Catalan de Charles V, Roi de France', *Bulletin de Géographie Histor. et Descr.*, 1895, 1.

Course, A.G., *A Dictionary of Nautical Terms*, Arco, London, 1962.

Davis, H.C., 'Records of Japanese Vessels driven upon the Northwest Coast of America', *Proceedings of the American Antiquarian Society*, 1872, 1,1.

Dimmock, L., 'The Chinese "Yuloh"' [Self-feathering propulsion oar], *Mariner's Mirror*, 1954, 40, 79.

Donnelly, I.A., *Chinese Junks and Other Native Craft*, Kelly and Walsh, Shanghai, 1924.

Donnelly, I.A., *Chinese Junks, a Book of Drawings in Black and White*, Kelly and Walsh, Shanghai, 1924.

Donnelly, I.A., 'River Craft of the Yangtzekiang', *Mariner's Mirror*, 1924, 10, 4.

Duyvendak, J.L.L., 'Sailing Directions of Chinese Voyages' (a Bodleian Library manuscript), *T'oung Pao* (*Archives concernant l'Histoire, les Langues, la Géographie, l'Ethnographie, et les Arts de l'Asie Orientale*, Leiden), 1938, 34, 230.

Duyvendak, J.L.L., *China's Discovery of Africa*, Probsthain, London, 1949. (Lectures given at London University, Jan. 1947.)

Duyvendak, J.L.L., 'The True Dates of the Chinese Maritime Expeditions in the Early Fifteenth Century', *Thung Pao* (as above), 1939, 34, 341.

Duyvendak, J.L.L., 'Ma Huan Re-examined', *Verhandelingen d. Koninklijke Akad. v. Wetenschappen te Amsterdam* (Afd. Letterkunde), 1933 (new series), 32, no. 3.

Elgar, F., 'Japanese Shipping', *Transactions (and Proceedings) of the Japan Society of London*, 1895, 3, 59.

Ferrand, G., *Relations de Voyages et Textes Géographiques Arabes, Persans, et Turcs relatifs à l'Extrême Orient, du 8ᵉ au 18ᵉ Siècles, traduites, revus et annotés etc.*, 2 vols, Leroux, Paris, 1913.

Ferrand, G., (tr.). *Voyage du Marchand Sulaymān en Inde et en Chine redigé en AD 851; suivi de remarques par Abñ Zayd Haṣan (vers AD 916)*, Bossard, Paris, 1922.

Fitzgerald, C.P., 'Boats of the Erh Hai Lake, Yunnan', *Mariner's Mirror*, 1943, 29, 135.

Gibson, C.E., *The Story of the Ship*, Schuman, New York, 1948; Abelard-Schuman, New York, 1958.

Gilfillan, S.C., *Inventing the Ship*, Follett, Chicago, 1935.

Goodrich, L. Carrington, *Short History of the Chinese People*, Harper, New York, 1943.

Hadi Hasan, *A History of Persian Navigation*, Methuen, London, 1928.

Hasler, H.G., 'Technically Interesting' [an account of the sailing-boat *Jester*, which finished second in the 1960 single-handed trans-Atlantic yacht race, rigged with a Chinese lug-sail], *Yachting World*, 1961, 113 (no. 2624), 14. See also 'Unusual Rig', *Yachting World*, 1958, 110 (no. 2589), 13.

Hejzlar, J., 'The Return of a Legendary Work of Art; the most famous Scroll in the Peking Palace Museum, "On the River during the Spring Festival" by Chang Tsê-Tuan (AD 1125)', *New Orient* (Prague), 1962, 3, (no. 1), 17.

Hennig, R., *Terrae Incognitae; eine Zusammenstellung und Kritische Bewertung der wichtigsten vorcolumbischen Entdeckungreisen an Hand der darüber vorliegenden Originalberichte*, 2nd ed., 4 vols., Brill, Leiden, 1944. (Includes most of the Chinese voyages of exploration, Chang Chhien [Zhang Qian], Kan Ying [Gang Ying], etc.

Hennig, R., 'Zur Frühgeschichte des Seeverkehrs im indischen Ozean', *Meereskunde* (Berlin), 1919, no. 151.

Herrmann, A., *Historical and Commercial Atlas of China*, Harvard-Yenching Institute, Cambridge, Mass., 1935.

Hirth, F., 'Uber den Seevekehr Chinas im Altertum nach chinesischen Quellen', *Geographische Zeitschrift*, 1896, 2, 444.

Hirth, F. and Rockhill, W.W., (tr.), *Chau Ju-Kua; His work on the Chinese and Arab Trade in the 12th and 13th centuries, entitled 'Chu-Fan-Chi'*, Imp. Acad. Sci., St Petersburg, 1911.

Holtzman, D., 'Shen Kua and his Mêng Chhi Pi Than'. *T'oung Pao (Archives concernant l'Histoire, les Langues, la Géographie, l'Ethnographie et les Arts de l'Asie Orientale*, Leiden), 1958, 46, 260.

Hornell, J., *Water Transport; Origins and Early Evolution*, Cambridge, 1946.

Hornell, J., 'Constructional Parallels in Scandinavian and Oceanic Boat Construction', *Mariner's Mirror*, 1935, 21, 411.

Hornell, J., 'The Origin of the Junk and Sampan', *Mariner's Mirror*, 1934, **20**, 331.

Hourani, G.F., *Arab Seafaring in the Indian Ocean in Ancient and Early Mediaeval Times*, Princeton Univ. Press, Princeton, N.J., 1951. (Princeton Oriental Studies, no. 13)

Hsiang Ta. 'A Great Chinese Navigator', *China Reconstructs*, 1965, **5**(no. 7), 11.

Hudson, G.F., *Europe and China: A Survey of their Relations from the Earliest Times to 1800*, Arnold, London, 1931.

Jal. A., *Archéologie Navale*, 2 vols., Arthus Bertrand, Paris, 1840.

Jal, A., *Glossaire Nautique; Repertoire Polyglot de Termes de Marine Anciennes et Modernes*. Didot, Paris, 1848.

Kirkman, J.S., 'Historical Archaeology in Kenya', *Antiquaries Journal*, 1957, **37**, 16.

Kirkman, J.S., *The Arab City of Gedi [near Malindi]; Excavations at the Great Mosque*, Oxford, 1954.

Kirkman, J.S., 'The Culture of the Kenya Coast in the Later Middle Ages; some Conclusions from Excavations 1948 to 1956', *South African Archaeological Bulletin*, 1956, **11**, 89.

Klaproth, J., *Lettre à M. le Baron A. de Humboldt, sur l'Invention de la Boussole*, Dondey-Dupré, Paris, 1934. (Résumés: P. de Laren-audière, *Bulletin de las Société de Géographie* (continued as *Le Géographie*), 1834, Oct; Anon, *Asiatic Journal and Monthly Register for British and Foreign India, China and Australia*, 1834 (2nd ser.), **15**, 105.

Kramer, J.B., 'The Early History of Magnetism', *Transactions of the Newcomen Society*, 1934, **14**, 183.

Kuwabara, Jitsuzo, 'On Phu Shou-Kêng, a man of the Western Regions, who was the Superintendent of the Trading Ships' Office in Chhüan-Chou towards the end of the Sung Dynasty, together with a general sketch of the Trade of the Arabs in China during the Thang and Sung eras', *Memoirs of the Research Department of Toyo Bunko* (Tokyo), 1928, **2**, 1; 1935, **7**, 1.

Landström, Björn, *The Ship; a Survey of the History of the Ship from the Primitive Raft to the Nuclear-Powered Submarine, with reconstructions in Words and Pictures*, Tr. from *Skeppet*, by M. Phillips, Allen & Unwin, London, 1961.

Layton, C.W.T., *A Dictionary of Nautical Words and Terms*, Brown & Ferguson, Glasgow, 1955.

Li Shu-Hua, 'Origine de la Boussole, II; Aimant et Boussole', *Isis*, 1954, **45**, 175. Engl. tr. with the addition of Chinese characters, *Chhing-Hua (T'sing Hua) Journal of Chinese Studies* (New Series, publ. Taiwan), 1956, **1** (no. 1), 81.

Li-Shu-Hua, 'Première Mention de l'Application de la Boussole à la Navigation.' *Oriens Extremus*, (Hamburg), 1954, **1**, 6.

Li Sun-Shêng, 'The Formosan Sea-going Raft and its Origin in Ancient China', *Bulletin of the Institute of Ethnology, Academia Sinica* (Taiwan), 1956, **1**, 25.

Lo Jung-Pang, 'The Emergence of China as a Sea-Power during the late Sung and early Yuan Periods', *Far Eastern Quarterly* (continued as *Journal of Asian Studies*), 1955, **14**, 489. Abstract in *Revue Bibliographique de Sinologie*, 1955, **1**, 66.

Lo Jung-Pang, 'The Decline of the Early Ming Navy', *Oriens Extremus* (Hamburg), 1958, **5**, 149.

Lo Jung-Pang, 'China's Paddle-Wheel Boats; the Mechanised Craft used in the Opium

War and their Historical Background', *Chhing-Hua (T'sing-Hua) Journal of Chinese Studies* (New Series, publ. Taiwan), 1960, 2 (no. 1), 189.

de Loture, R & Haffner, L., *La Navigation à travers les Ages; Evolution de la Technique Nautique et de ses Applications*, Payot, Paris, 1952.

Lovegrove, H., 'Junks of the Canton River and the West River System', *Mariner's Mirror*, 1932, 18, 241.

McGregor, J., 'On the Paddle-Wheel and Screw Propeller, from the Earliest Times', *Journal of the Royal Society of Arts*, 1858, 6, 335.

McRobert, I., 'The Chinese Yuloh' [self-feathering propulsion oar], *Mariner's Mirror*, 1940, 26, 313.

Mathew, G., 'The Culture of the East African Coast in the Seventeenth and Eighteenth Centuries, in the Light of recent Archaeological Discoveries', *Man*, 1956, 56, 65.

Mathew, G., 'Chinese Porcelain in East Africa and on the Coast of South Arabia', *Oriental Art*, 1956 (n.s.), 2, 50.

May, W.E., 'Historical Notes on the Deviation of the Compass', *Terrestrial Magnetism and Atmospheric Electricity* (continued as *Journal of Geophysical Research*), 1947, 217.

May, W.E., 'The History of the Magnetic Compass', *Mariner's Mirror*, 1952, 38, 210.

May, W.E., 'The Birth of the Compass', *Journal of the Institute of Navigation*, 1949, 2, 259.

May, W.E., 'Alexander Neckham (*c.* AD 1187) and the Pivoted Compass Needle', *Journal of the Institute of Navigation*, 1955, 8, 283.

Mayers, W.F., 'Chinese Explorations of the Indian Ocean during the fifteenth century AD', (Partly a translation of the *Hsi-Yang Chhao Kung Tien Lu [Xi-Yang Chao Gong Dian Lu]* of Huang Shêng-Tsêng [Huang Sheng-Zeng], AD 1520.) *China Review* (Hongkong and Shanghai), 1875, 3, 219, 331; 1875, 4, 61, 173.

Mayers, W.F., 'Chinese Junk Building', *Notes and Queries on China and Japan*, 1867, 1, 170.

Mills, J.V., 'Malaya in the *Wu Pei Chih* Charts', *Journal of the Malaysian Branch of the Royal Asiatic Society*, 1937, 15 (no. 3), 1.

Mills, J.V., 'Notes on Early Chinese Voyages', *Journal of the Royal Asiatic Society*, 1951, 17.

Mills, J.V., 'The Largest Chinese Junk and its Displacement', *Mariner's Mirror*, 1960, 46, 147.

Mitchell, A. Crichton, 'Chapters in the History of Terrestrial Magnetism [I. The Discovery of Directivity]', *Terrestrial Magnetism and Atmospheric Electricity* (continued as *Journal of Geophysical Research*), 1932, 37, 105.

Mitchell, A. Crichton, Chapters in the History of Terrestrial Magnetism [II. The Discovery of Declination]', *Terrestrial Magnetism and Atmospheric Electricity*, 1937, 42, 241.

Mitchell, A. Crichton, 'Chapters in the History of Terrestrial Magnetism [III. The Discovery of Dip]', *Terrestrial Magnetism and Atmospheric Electricity*, 1939, 44, 77.

Mitchell, A. Crichton, 'Chapters in the History of Terrestrial Magnetism [IV. The Development of Magnetic Science in Classical Antiquity]', *Terrestrial Magnetism and Atmospheric Electricity*, 1946, 51, 323.

Moll, F., *Das Schiff in der bildenden Kunst vom Altertum bis zum Ausgang des Mittelal-ters*, Schroeder, Bonn, 1929.

Moll, F., 'History of the Anchor', *Mariner's Mirror*, 1927, 13, 293. 'Die Entwicklung des Schiffsankers bis zum Jahre 1500 n. Chr', *Beiträge z. Gesch. d. Technik u. Industrie* (continued as *Technik Geschichte*), 1919, 9, 41.

Moll, F. and Laughton, L.G. Carr, 'The Navy of the province of Fukein', *Mariner's Mirror*, 1923, 9, 364.

Mukerji, Radhakamud, *Indian Shipping; a History of the Sea-Borne Trade and Maritime Activity of the Indians from the Earliest Times*, Longmans Green, Bombay and Calcutta, 1912.

Mulder, W.Z., 'The *Wu Pei Chih* Charts', *T'oung Pao* (*Archives concernant l'Histoire, les Langues, la Géographie, l'Ethnographie et les Arts de l'Asie Orientale*, Leiden), 1944, 37, 1.

Needham, Joseph, 'The Chinese Contributions to the Development of the Mariner's Compass', Abstract in *Resumo das Communicações do Congresso Internacional de História dos Descobrimentos*, p. 273. Lisbon, 1960. *Actas*, Lisbon, 1961, 2, 311. Also *Scientia*, 1961, 96, 225.

Needham, Joseph, 'The Chinese Contributions to Vessel Control', Abstract in *Resumo das Communicações do Congresso Internacional de História dos Descobrimentos*, Lisbon, 1960, 274. *Actas*, Lisbon, 1961, 2, 235. Also *Scientia*, 1961, 96, 123, 163. Polish abridgement by W.A. Drapella, *Bulletyn Nutologyczny* (Gdynia), 1963–4, 6–7, 33. And (with illustrations), French translation as article in 'Les Aspects Internationaux de la Découverte Océanique aux 15ᵉ et 16ᵉ Siècles', *Actes de Vᵉ Colloque International d'Histoire Maritime*, Lisbon, 1960, Sevpen, Paris, 1966.

des Noëttes, R.J.E.C. Lefebvre, *De la Marine Antique à la Marine Moderne; La Révolution du Gouvernail*, Masson, Paris, 1935.

des Noëttes, R.J.E.C. Lefebvre, 'Le Gouvernail; contribution à l'Histoire de l'Esclavage', *Mémoires de la Société (Nat.) des Antiquaires de France*, 1932 (8ᵉ ser.), 8 (78), 24.

des Noëttes, R.J.E.C. Lefebvre, 'Autour de Vaisseau de Borobodur; l'Invention du Gouvernail', *La Nature*, 1932, i; 1934 (no. 2934, 1 Aug.), 97.

van Nouhuys, J.W. 'Chinese Anchors', *Mariner's Mirror*, 1925, 11, 96.

Paik, L.G., 'From Koryu to Kyung by Soh Keung, Imperial Chinese Envoy to Korea in AD 1124' (excerpts concerning boats and ships from Hsü Ching's [Xu Jing's] *Kao-Li Thu Chìng [Gao-Li Tu Jing]*), Printed as appendix to Underwood, H.H., *Journal* (or *Transactions*) *of the Korea Branch of the Royal Asiatic Society*, 1933, 23, 90.

Pao Tsun-Phêng, *On the Ships of Chêng-Ho*, Chung-Hua Tshung-Shu [Zhong-Hua Cong-Shu], Thaipei [Daibai] and Hongkong, 1961. (National Historical Museum Collected Papers on the History and Art of China, 1st ser., no. 6.)

Pao Tsun-Phêng, *A Study of the 'Castled Ships' of the Han Period*, National Historical Museum, Thaipei [Daibai], Taiwan, 1967. (Collected Papers on the History and Art of China, no. 2.)

Paris, F.E. Admiral, *Essai sur la Construction Navales des Peuples Extra-Européens*, Arthus Bertrand, Paris, no date (1841–3).

Paris, F.E. Admiral, *Souvenirs de Marine; Collection de Plans ou Dessins de Navires et de*

Bateaux Anciens ou Modernes, Existants ou Disparus, Gauthier-Villars, Paris. Vol. I, 1882, contains Japanese and Indo-Chinese material, vol. II, 1884, contains some Chinese material, vol. III, 1886, vol. IV, 1889 contains some Chinese material, vol. V, 1892, vol. VI, 1908, contains some Japanese material.

Pelliot, P., 'Les Grandes Voyages Maritimes Chinois au Début du 15ᵉ Siècle' (review of Duyvendak 'Ma Huan re-examined' [see above]), *T'oung Pao (Archives concerncnt l'Histoire, les Langues, la Géographie, l'Ethnographie et les Arts Orientale*, Leiden), 1933, 30, 237. Chinese translation by Fêng Chhêng-Chün [Feng Cheng-Jun], Shanghai, 1935, entitled *Chêng Ho Hsia Hsi-Yang Khao [Cheng Ho Xia Xi-Yang Kao]*.

Pelliot, P., 'Notes additionelles sur Tcheng Houo (Chêng-Ho [Zheng-Ho]) et sur ses Voyages', *T'oung Pao*, 1934, 31, 210.

Pelliot, P., 'Encore à Propos des Voyages de Tcheng Houo (Chêng-Ho [Zheng-Ho])', *T'oung Pao*, 1936, 32, 210.

Peri, N., 'A Propos du Mot "Sampan"', *Bulletin de l'Ecole Française de l'Extrême Orient* (Hanoi), 1919, 19 (no. 5), 13.

P [errin], W.G., 'The Balanced Rudder', *Mariner's Mirror*, 1926, 12, 232.

Petersen, E. Allen, *In a Junk across the Pacific*, Elek, London, 1954.

Phillips, G., 'The Seaports of India and Ceylon, described by Chinese Voyagers of the Fifteenth Century, together with an account of Chinese Navigation', *Journal (or Transactions) of the North China Branch of the Royal Asiatic Society*, 1885, 20, 209; 1886, 21, 30 (both with large folding maps).

Playfair, G.M.H., 'Watertight Compartments in Chinese Vessels', *Journal (or Transactions) of the North China Branch of the Royal Asiatic Society*, 1886, 21, 106. (Quotation of a letter of Benjamin Franklin.)

Poujade, J., *La Route des Indes et Ses Navires*, Payot, Paris, 1946.

Poujade, J., *Les Jonques des Chinois du Siam* (in relation to the ship sculptured on the Bayon of the Angkor Vat), Publication du Centre de Research Culturelle de la Route des Indes, Gauthier-Villars, Paris, 1946. (Documents d'Ethnographie Navale, Fasc. 1.)

Pritchard, L.A., *The 'Ningpo' Junk* (voyage from Shanghai to San Pedro, California, 1912/13), *Mariner's Mirror*, 1923, 9, 89 and notes by H. Sz[ymanski], p. 312 and G.A. B[allard], p. 316.

Purvis, F.P., 'Ship Construction in Japan', *Transactions of the Asiatic Society of Japan*, 1919, 47, 1; 'Japanese Ships of the Past and Present.' *Transactions (and Proceedings) of the Japan Society of London*, 1925, 23, 51.

Reischauer, E.O., 'Notes on the Thang Dynasty Sea-Routes', *Harvard Journal of Asiatic Studies*, 1940, 5, 142.

la Roërie, G., 'l'Histoire de Gouvernail', *Revue Maritime*, 1938 (no. 219), 309; (no. 220), 481. Also separately, Soc. d'Editions Géeographiques Maritimes et Coloniales, Paris, 1938.

la Roërie, G. &Vivielle, J., *Navires et Marins, de la Rame à Hélice*, 2 vols. Ducharte & van Buggenhoudt, Paris, 1930.

Schoff, W.H., 'Navigation to the Far East under the Roman Empire', *Journal of the American Oriental Society*, 1917, 37, 240.

Schück, [K.W.] A. *Der Kompass*. 2 vols, Privately printed Hamburg, 1911–1915. (The second volume, *Sagen von der Erfindung des Kompasses; Magnet, Calamita*,

Bussole, Kompass; Die Vorgänger des Kompasses, contains a good deal on the Chinese material, in so far as it could be evaluated at the time, mainly on a long account of 18th- and 19th-century European views about it. This had seen preliminary publication in *Die Natur* (Halle a/d Salle), 1891, 40, (nos. 51 and 52).)

Smith, P.J. & Needham, Joseph, 'Magnetic Declination in Mediaeval China', *Nature*, 1967, **214**, 1213.

Smyth, H. Warington, *Mast and Sail in Europe and Asia*, Blackwood, Edinburgh, 1906; 2nd ed. 1929.

Spratt, H.P., *The Birth of the Steamboat*, Griffin, London, 1959.

Stevenson, E.L., *Terrestrial and Celestial Globes; their History and Construction ...*, 2 vols., Hispanic Society of America (Yale University Press), New Haven, 1921.

Stevenson, E.L., *Portolan Charts; their Origin and Characteristics ...*, Hispanic Society of America, New York, 1911.

Sullivan, M., 'Chinese Export Porcelain in Singapore', *Oriental Art*, 1957, **3**, 145.

Sun Jen I-Tu & Sun Hsüeh-Chuan, *'Thien Kung Khai Wu'*, *Chinese Technology in the Seventeenth Century, by Sung Ying-Hsing*, Pennsylvania State University Press; University Park and London, Penn., 1966.

Tarn, W.W., *Hellenistic Military and Naval Developments*, Cambridge, 1930. (Lees-Knowles Lectures in Military History, Cambridge.)

Taylor, E.G.R., 'The South-Pointing Needle', *Imago Mundi: Yearbook of Early Cartography*, 1951, **8**, 1.

Taylor, E.G.R., *The Haven-Finding Art; a History of Navigation from Odysseus to Captain Cook*, Hollis & Carter, London, 1956.

Thompson, Sylvanus P., *Peregrinus and his 'Epistola'*, London, 1907.

Ting Wên-Chiang & Donnelly, I.A. (tr.), ' "Things Produced by the Works of Nature", published 1639, translated by Dr V.K. Ting', *Mariner's Mirror*, 1925, **11**, 234. (A translation, with notes, of that part of chapter 9 of *Thien Kung Khai Wu* [*Tian Gong Kai Wu*] (AD 1637) which deals with nautical technology.)

Torr, C., *Ancient Ships*, Cambridge, 1894. Reprinted 1964, edited and introduced by A.J. Podlecki, with an appendix containing a series of articles on the Greek warship and the Greek trireme, by W.W. Tarn, A.B. Cook, C. Torr, W. Richardson & P.H. Newman; and many new illustrations (Argonaut, Chicago).

Underwood, H.H., 'Korean Boats and Ships', *Journal* (or *Transactions*) *of the Korea Branch of the Royal Asiatic Society*, 1933, **23**, 1–100.

Villiers, A., *Sons of Sinbad; an Account of sailing with the Arabs in their Dhows, in the Red Sea, round the Coasts of Arabia and to Zanzibar and Tanganyika; Pearling in the Persian Gulf; and the Life of the Shipmasters and Mariners of Kuwait*, Hodder & Stoughton, London, 1940.

Villiers, A., *The Indian Ocean*, Museum, London, 1952.

Villiers, A., *Monsoon Seas; the Story of the Indian Ocean*, McGraw-Hill, New York, 1952.

Wang Kung-Wu, 'The Nanhai Trade; a Study of the Early History of Chinese Trade in the South China Sea [from later Han to Wu Tai [Wu Dai] 1st to 10th centuries AD]', *Journal of the Malayan Branch of the Royal Asiatic Society*, 1958, **31** (part 2), 1–135.

Waters, D.W., 'Chinese Junks; the Antung Trader', *Mariner's Mirror*, 1938, **24**, 49.

Waters, D.W., 'Chinese Junks; the Pechili Trader', *Mariner's Mirror*, 1939, 25, 62.

Waters, D.W., 'Chinese Junks, an Exception; the Tongkung', *Mariner's Mirror*, 1940, 26, 79.

Waters, D.W., 'Chinese Junks; the Twaqo', *Mariner's Mirror*, 1946, 32, 155.

Waters, D.W., 'Chinese Junks; the Hangchow Bay Trader and Fisher', *Mariner's Mirror*, 1947, 33, 28.

Waters, D.W., 'The Chinese Yuloh' [self-feathering propulsion oar], *Mariner's Mirror*, 1946, 32, 189.

Waters, D.W., 'The Straight, and other Chinese "Yulohs"'. *Mariner's Mirror*, 1955, 41, 60.

Waters, D.W., 'Some Coastal Sampans of North China: I. The Sampan of the Antung Trader', Manuscript Notes on Models in the National Maritime Museum, Greenwich, the Science Museum, South Kensington, and the Mystic Seaport Museum, Mystic, Conn., U.S.A.

Waters, D.W., 'Some Coastal Sampans of North China: II. The "Duck", "Chicken" and "Open Bow" Sampans of Wei-Hai-Wei', Manuscript Notes as above.

Worcester, G.R.G., *Junks and Sampans of the Upper Yangtze*, Inspector-General of Customs, Shanghai, 1940. (China Maritime Customs Pub., ser. III, Miscellaneous, no. 51.)

Worcester, G.R.G., *Notes on the Crooked-Bow and Crooked-Stern Junks of Szechuan*, Inspectorate-General of Customs, Shanghai, 1941. (China Maritime Customs Pub., ser. III, Miscellaneous, no. 52.)

Worcester, G.R.G., *The Junks and Sampans of the Yangtze; a study in Chinese Nautical Research*: Vol. 1, *Introduction, and Craft of the Estuary and Shanghai Area*; Vol. 2, *The Craft of the Lower and Middle Yangtze and Tributaries*, Inspectorate-General of Customs, Shanghai, 1947, 1948. (China Maritime Customs Pub., ser. III, nos. 53, 54.)

Worcester, G.R.G., 'The Chinese War-Junk', *Mariner's Mirror*, 1948, 34, 16.

Worcester, G.R.G., 'The Origin and Observance of the Dragon-Boat Festival in China', *Mariner's Mirror*, 1956, 42, 127.

Worcester, G.R.G. 'Four Small Craft of Taiwan', *Mariner's Mirror*, 1956, 42, 302.

Worcester, G.R.G., 'The Inflated Skin-Rafts of the Huang Ho', *Mariner's Mirror*, 1957, 43, 73.

Worcester, G.R.G., 'Six Craft of Kuangtung', *Mariner's Mirror*, 1959, 45, 130.

Worcester, G.R.G., *Sail and Sweep in China; the History and Development of the Chinese Junk as Illustrated by the Collection of Models in the Science Museum* [*London*], Her Majesty's Stationery Office, London, 1966.

Yule, Sir Henry, *Cathay and the Way Thither; being a Collection of Mediaeval Notices of China*, Hakluyt Society Publications 2nd ser.) London, 1913–15. (1st edition 1866.) Revised by H. Cordier, 4 vols. Vol. 1 (no. 38), *Introduction; Preliminary Essay on the Intercourse between China and the Western Nations previous to the Discovery of the Cape Route*; Vol 2 (no. 33), *Odoric of Pordenone*; Vol. 3 (no. 37), *John of Monte Corvino and Others*; Vol. 4 (no. 41), *Ibn Battutah and Benedict of Goes*. (Photographic reprint, Peking, 1942.)

INDEX

Abundance of Jottings by Old Mr (Li)
Chieh-An, see *Chieh An Lao Jen
Man Pi*
Account of the Liu-Chhiu Islands, see
Liu-Chhiu Kuo Chin Lüeh
action at a distance, and magnetism, in
China, 5
Adelard of Bath, 9
Aden, Gulf of, Chinese description of,
133
Admirals' titles
Admiral of the Triple Treasure. *See*
Chêng Ho
Prancing-Dragon Admiral. *See* Wang
Chün
Three-Jewel Eunuch. *See.*Chêng Ho
Torrent-Descending Commander,
102, 103
Africa,
Arab trading centres in, 132, 133
Chinese map of southern region of,
136
Chinese relationships with, 132–9
de Albuquerque, Alfonso, 142,.152
alchemists and magnetism,
in the Han, 20
in medieval China, 27
'Alexander Romance', 48
and diving, 250
Alexandria,
dry dock at, 240
lighthouse (Pharos) at, 241
Alfonso X (King of Castile), 177
Algol, and image-chess, 51
amber, 5
preparing imitations of, 5, 6

electrostatic test for, 6
America, Pre-Columbian, and the
Chinese, 153–9
Amerindian civilisations, and possible
contact with Chinese, 154–6
Analysis of the Magnetic Compass, see
Lo Ching Chieh
Analytical Dictionary of Characters, see
Shou Wên
anchors,
adze type, 237, 238
– stock on, 237
ancient Egyptian, 236
bronze, Homeric, 236
Chinese, 236–9
on Han tomb ship model, 239
hawsers for, 237, 238
metal hooks on, 236
parts of, 236
sea type, Chinese, 237, 238
'ship's self' type, Chinese, 238
'watchdog' type, 238
'water-eye' type, Chinese, 239
windlasses for, 111, 237
Angkor Thom, Cambodia, carving of
Chinese merchant ship at, 111–13,
227
Apparatus for Determining the Sun's
Position, see *Khuei Jih Chhi*
Arabs,
appearance of magnetic compass
among, 9
medieval trading centres of in Africa,
132, 133
Archimedes, 3
Aristotle, 3